A Few Seconds to Midnight

Andrew Parry

Published by Andrew Parry, 2024.

A FEW SECONDS TO MIDNIGHT

First edition. September 27, 2024.

Copyright © 2024 Andrew Parry.

ISBN: 979-8227750419

Written by Andrew Parry.

Table of Contents

Introduction: A Few Seconds to Midnight

In the year 2024, the world teeters on the brink of a catastrophe that many believed had been relegated to the annals of history. The ominous specter of nuclear war, once the grim preoccupation of the Cold War era, has resurfaced with terrifying urgency. The Doomsday Clock, maintained by the Bulletin of the Atomic Scientists, now stands perilously close to midnight, symbolizing humanity's proximity to self-annihilation. This is not merely a symbolic gesture; it reflects the harsh reality that we are now closer to nuclear conflict than ever before.

The current geopolitical landscape is fraught with tension, with Russia at the center of this global crisis. President Vladimir Putin, a leader known for his strategic maneuvering and defiance of Western influence, has issued stark warnings that echo the darkest days of the 20th century. His rhetoric, coupled with a series of provocative military actions, suggests that the unthinkable—a full-scale nuclear war—may be more than just a threat. It could be an imminent reality.

Russia's aggression is not happening in a vacuum. The world is witnessing a dangerous escalation of conflicts that have long simmered beneath the surface. From the ongoing tensions in Ukraine, where Russian forces have made significant and brutal incursions, to the broader struggles for power in Eastern Europe and beyond, the potential for these conflicts to spiral out of control is terrifyingly real. Putin's warnings, coupled with the military posturing and tests of nuclear-capable missiles, have brought the world to the edge of a precipice.

This book is born out of this dire situation. It is not merely a guide to constructing a physical structure but a comprehensive manual for survival in an age where nuclear war is no longer a distant possibility but a looming threat. As the world grapples with the resurgence of nuclear brinkmanship, the need for preparedness has never been more critical.

The aim of this book is to equip you with the knowledge and tools necessary to protect yourself and your loved ones should the worst come to pass. It will guide you through the technical and scientific aspects of nuclear fallout, helping you understand the dangers of radiation, the protective capabilities of various building materials, and the medical treatments required for radiation exposure. You will learn how to design and build a fallout shelter tailored to your specific needs and location, ensuring maximum protection against the devastating effects of nuclear war.

But this book is about more than just the immediate aftermath of a nuclear attack. It addresses the long-term challenges of surviving in a post-apocalyptic world, where the threat of nuclear winter, food and water contamination, and societal collapse are very real. It offers strategies for long-term survival, from stockpiling essential supplies to cultivating food in low-light conditions and maintaining physical and mental health in the face of prolonged adversity.

In these pages, you will find detailed instructions, practical advice, and a wealth of knowledge gathered from experts in the fields of civil defense, emergency management, and survival. The information presented here is designed to be accessible and actionable, enabling you to take the necessary steps to safeguard your future.

As we stand at a few seconds to midnight, it is crucial to acknowledge the gravity of the situation. While we hope that diplomacy and reason will prevail, the reality is that we must prepare for every possibility. This book is a response to that necessity—a resource for those who choose to take their safety into their own hands in an increasingly uncertain world.

Prepare yourself, for the world we once knew may be on the brink of irreversible change. In the shadow of nuclear war, knowledge and preparation are your most powerful allies. This book will be your guide through the darkness, offering you the means to survive and endure when the unthinkable becomes reality.

Understanding Nuclear Fallout: The Basics

When we talk about nuclear fallout, we're diving into one of the most serious and life-altering consequences of a nuclear explosion. To truly understand the gravity of nuclear fallout, we first need to break down what it is, how it forms, and why it's so dangerous. This chapter will guide you through the basics, laying the foundation for everything that follows.

Nuclear fallout refers to the residual radioactive material propelled into the upper atmosphere following a nuclear explosion. This material "falls out" of the sky after the explosion and the vaporized matter from the weapon itself cools and condenses into solid particles. These particles, which can range from microscopic dust to larger fragments, are radioactive and pose a severe risk to health and the environment.

The radioactive particles in fallout are the by-products of nuclear fission—the process where the nucleus of an atom splits into smaller parts, releasing an enormous amount of energy. In a nuclear explosion, this energy is unleashed in the form of an intense blast, heat, and radiation. The immediate effects of the explosion, like the blast and heat, cause widespread destruction, but it's the lingering radiation in the form of fallout that poses a long-term threat.

Fallout is composed of various radioactive isotopes, with some of the most concerning being cesium-137, iodine-131, and strontium-90. These isotopes emit radiation in the form of alpha, beta, and gamma rays. Alpha particles are the least penetrating, but if inhaled or ingested, they can cause significant internal damage. Beta particles are more penetrating and can cause skin burns and internal harm if ingested. Gamma rays are the most dangerous, capable of penetrating deep into the body and causing widespread tissue damage.

One of the most critical aspects of understanding fallout is recognizing how it spreads and what influences its distribution. The explosion's altitude, the yield of the weapon, and the weather conditions at the time all play a role. For instance, a ground-level explosion generates more fallout than an airburst because it pulls in more earth and debris that become irradiated and then carried by the wind. Wind patterns can transport fallout particles across vast distances, affecting areas far removed from the explosion site.

The danger of fallout lies not just in its immediate effects but in its persistence. Radioactive particles remain hazardous for varying periods, depending on their half-lives—the time it takes for half of the radioactive atoms to decay. Some isotopes, like iodine-131, have relatively short half-lives (around 8 days) but are still dangerous due to their rapid uptake by the human thyroid gland. Others, like cesium-137, have longer half-lives (about 30 years), posing a prolonged environmental threat.

Exposure to fallout can occur through different pathways: inhalation, ingestion, or direct contact with the skin. Inhalation is particularly dangerous because radioactive particles can lodge in the lungs, leading to severe internal radiation exposure. Ingestion of contaminated food or water can introduce radioactive materials into the body, where they can accumulate in various organs, causing damage over time. Direct contact with fallout particles can cause radiation burns on the skin and lead to an increased risk of cancer.

Understanding the health effects of radiation exposure is crucial. In the short term, high doses of radiation can cause acute radiation sickness, characterized by symptoms like nausea, vomiting, diarrhoea, and in severe cases, death. Long-term exposure, even to lower levels of radiation, increases the risk of developing cancers, particularly leukemia, and can cause genetic mutations that affect future generations.

Another aspect to consider is the environmental impact of fallout. Radioactive contamination can make areas uninhabitable for years, affecting soil, water, and ecosystems. Plants and animals in contaminated areas can accumulate radioactive materials, leading to a bioaccumulation effect, where radiation concentrations increase as you move up the food chain. This contamination can severely impact agriculture, making food supplies unsafe for consumption.

As we delve deeper into the technical and scientific aspects of nuclear fallout in subsequent chapters, keep in mind the basics we've covered here. Understanding what fallout is, how it forms, and why it's dangerous will provide a solid foundation as we explore the specifics of how to protect yourself and your family, the materials that can shield you from radiation, and the best practices for building a fallout shelter.

In the event of a nuclear explosion, the immediate danger is obvious, but it's the invisible, lingering threat of fallout that requires careful preparation and understanding. This chapter has set the stage for that understanding, providing you with the essential knowledge to grasp the complexity of nuclear fallout and the urgency of proper protection measures.

The Science behind Radioactive Decay

Radioactive decay is the fundamental process that underpins the threat posed by nuclear fallout. To fully comprehend the dangers and how best to protect ourselves, we must first understand what radioactive decay is, how it works, and its implications for both short-term and long-term survival. This chapter will explore the science behind radioactive decay, explaining the key concepts in a way that's accessible and relevant to the task of building a nuclear fallout shelter.

At its core, radioactive decay is the process by which an unstable atomic nucleus loses energy by emitting radiation. This process is spontaneous and occurs naturally as certain isotopes, known as radionuclides, seek stability. These radionuclides are the by-products of nuclear reactions, such as those occurring in a nuclear explosion, where the immense energy released leads to the formation of a variety of unstable isotopes.

Radioactive decay can happen in several ways, but the most common types of decay relevant to fallout are alpha decay, beta decay, and gamma decay. Each type of decay involves the emission of different particles or electromagnetic waves, which have varying levels of energy and penetration power, and thus different implications for human health and safety.

Alpha decay occurs when an unstable nucleus emits an alpha particle, which is composed of two protons and two neutrons—essentially a helium nucleus. Alpha particles are relatively large and carry a positive charge, but they have low penetration power. They can be stopped by something as thin as a sheet of paper or even the outer layer of human skin. However, the danger arises if alpha-emitting materials are inhaled, ingested, or enter the body through a wound. Once inside the body, alpha particles can cause significant cellular damage due to their high energy, leading to increased risks of cancer and other serious health issues.

Beta decay involves the emission of a beta particle, which can be either an electron or a positron, depending on whether the decay is beta-minus or beta-plus. Beta particles are smaller than alpha particles and carry either a negative or positive charge. They have greater penetration power than alpha particles, capable of passing through skin and potentially causing radiation burns or more significant internal damage if they penetrate deeply enough. Beta radiation is particularly concerning for fallout because it can contaminate the skin, food, and water, leading to both external and internal exposure.

Gamma decay is different from alpha and beta decay in that it involves the emission of gamma rays, which are high-energy photons, rather than particles. Gamma rays have no mass and no charge, but they are highly penetrating, capable of passing through most materials, including human tissue, with ease. Gamma radiation is the most dangerous form of radiation associated with fallout because it can cause extensive internal damage without any direct contact. Shielding against gamma rays requires dense materials, such as lead or thick layers of concrete, making it a critical consideration in the design and construction of a fallout shelter.

A key concept in radioactive decay is the half-life of a radionuclide, which is the time it takes for half of the atoms in a given sample to decay into a more stable form. The half-life varies widely among different isotopes, ranging from fractions of a second to millions of years. For example, iodine-131, a common isotope in fallout, has a half-life of about 8 days, which means it remains dangerous for weeks after a nuclear explosion. On the other hand, cesium-137, with a half-life of around 30 years, remains hazardous for decades, posing long-term environmental and health risks.

Understanding half-lives is crucial because it informs how long certain areas may remain unsafe after a nuclear event and how long protective measures, such as sheltering or avoiding contaminated food and water, need to be maintained. Radionuclides with short half-lives may pose intense but short-term risks, while those with longer half-lives require long-term strategies for mitigation and safety.

Another important concept is the decay chain, which is the series of successive radioactive decays that certain radionuclides undergo as they move toward a stable state. For example, uranium-238 decays through a complex chain of transformations, eventually becoming lead-206, a stable isotope. Each step in the decay chain may produce different types of radiation, contributing to the complexity and duration of the radioactive hazard in fallout. Understanding these chains is essential for predicting the behavior and risks associated with specific fallout scenarios.

Radioactive decay is not only a source of danger but also a tool for scientists to measure time and contamination levels. Devices like Geiger counters and scintillation detectors measure the rate of decay, providing critical information about the intensity and distribution of fallout. This data can be used to map out safe zones, determine the necessity of evacuation, or plan the duration of sheltering in place.

In the context of a nuclear fallout shelter, the science of radioactive decay dictates the materials you choose for construction, the design of the shelter, and the length of time you need to remain inside. For instance, understanding that gamma rays are highly penetrating leads to the decision to use dense materials like concrete or lead for shielding. Knowing the half-lives of various radionuclides informs how long supplies must last and when it might be safe to emerge.

Moreover, radioactive decay has implications beyond the immediate aftermath of a nuclear event. Long-term exposure, even at low levels, can lead to chronic health issues, environmental degradation, and challenges in agricultural production.

This ongoing risk necessitates a comprehensive approach to fallout shelter design and preparedness, one that considers not only survival in the immediate aftermath but also the ability to thrive in a world where radioactive contamination may persist for years or even decades.

As you build your understanding of radioactive decay, keep in mind that this knowledge is not just academic; it is practical and vital. It informs every decision you make in preparing for nuclear fallout, from the materials you choose to the way you structure your survival plans. By mastering the science of radioactive decay, you equip yourself with the tools needed to protect yourself and your loved ones in one of the most extreme survival scenarios imaginable.

Short-Term Radiation Exposure: Immediate Dangers

In the immediate aftermath of a nuclear explosion, one of the most pressing concerns is short-term radiation exposure. Understanding the dangers that accompany this exposure is crucial for taking the right steps to protect yourself and your loved ones. This chapter delves into the immediate risks associated with short-term radiation exposure, what to expect, and how to respond effectively.

Short-term radiation exposure, often referred to as acute radiation exposure, occurs when the body is subjected to a high dose of radiation over a brief period. This type of exposure is typically experienced in the minutes, hours, or days following a nuclear detonation when radioactive materials are most concentrated in the environment. The severity of the effects depends on the dose of radiation received, which is measured in units called Sieverts (Sv). The higher the dose, the more severe the health effects.

One of the first things to understand about short-term radiation exposure is how quickly it can become life-threatening. Exposure to extremely high levels of radiation—above 10 sieverts, for example—can cause immediate and severe damage to the body's tissues and organs. This is because radiation disrupts the molecular structure of cells, particularly the DNA, leading to cell death or malfunction. The most radiation-sensitive cells in the body are those that divide rapidly, such as those in the bone marrow, gastrointestinal tract, and reproductive organs.

The most immediate health consequence of acute radiation exposure is Acute Radiation Syndrome (ARS), also known as radiation sickness. ARS occurs in stages, with symptoms appearing in a predictable pattern based on the level of exposure. The first phase is the prodromal stage, which begins within hours of exposure. During this stage, symptoms such as nausea, vomiting, diarrhea, and fatigue manifest. These symptoms can be severe, depending on the radiation dose, and are often mistaken for other illnesses, which can delay appropriate treatment.

Following the prodromal stage, there may be a brief period where symptoms seem to subside. This latent stage can last from a few hours to several days, depending on the radiation dose. However, this apparent recovery is misleading, as the body is undergoing significant internal damage, particularly to the bone marrow, which is critical for producing blood cells.

The third stage of ARS is the manifest illness stage, where the full effects of radiation damage become apparent. Symptoms during this stage can include severe gastrointestinal issues, such as uncontrollable diarrhoea and bleeding, which can lead to dehydration and electrolyte imbalance. Damage to the bone marrow can result in a drastic reduction in white blood cells, leading to a weakened immune system and making the body highly susceptible to infections. Additionally, a significant drop in platelets increases the risk of uncontrolled bleeding, both internally and externally.

At high doses of radiation, damage to the central nervous system becomes a critical concern. Symptoms can include dizziness, disorientation, loss of consciousness, and even seizures. Central nervous system damage is often fatal within hours to days, as the body's ability to regulate vital functions becomes compromised.

In cases of extreme radiation exposure, death can occur within days or even hours. This is typically due to the destruction of the body's vital systems, particularly the gastrointestinal tract and central nervous system. At slightly lower, but still high doses, death may occur within weeks due to complications from infections, bleeding, or other organ failures as a result of the cumulative damage caused by radiation.

For those who survive the initial acute effects of radiation exposure, the long-term consequences can be severe. Survivors of ARS may face an increased risk of developing cancers, particularly leukemia, due to the damage inflicted on the DNA. The risk of other chronic conditions, such as cardiovascular disease, also increases. Furthermore, damage to the reproductive organs can lead to infertility or, in some cases, genetic mutations that can affect future offspring.

Given the severe and often fatal consequences of short-term radiation exposure, immediate protective measures are essential following a nuclear event. The most effective way to protect against radiation exposure is to limit your time in high-radiation areas, increase your distance from the source of radiation, and use appropriate shielding. In the context of a nuclear fallout shelter, this means getting to your shelter as quickly as possible, staying inside until it is safe to emerge, and ensuring that your shelter is constructed from materials that provide adequate protection from radiation.

If exposure does occur, it's crucial to decontaminate as soon as possible. Removing contaminated clothing and washing your body thoroughly can reduce the amount of radiation your body absorbs. Seeking medical attention immediately is also essential, as early intervention can mitigate some of the severe effects of radiation exposure. Treatments for ARS are primarily supportive and include measures to manage symptoms, prevent infections, and, in some cases, stimulate the recovery of bone marrow function.

Another key aspect of responding to short-term radiation exposure is monitoring and measuring the dose received. Personal dosimeters, Geiger counters, and other radiation detection devices can provide critical information about your level of exposure, allowing you to make informed decisions about when it's safe to exit your shelter or seek medical help.

In addition to personal safety, understanding the dangers of short-term radiation exposure is vital for community planning and emergency response. Local authorities and emergency services need to be prepared to provide timely information and assistance, including evacuation orders, shelter locations, and medical treatment facilities equipped to handle radiation injuries.

In summary, short-term radiation exposure poses immediate and severe health risks, including the potential for Acute Radiation Syndrome, which can be fatal without prompt and effective intervention. Protecting yourself from these dangers requires quick action, proper sheltering, and an understanding of the symptoms and treatments associated with radiation exposure. As you continue to prepare for the possibility of a nuclear event, this knowledge will be crucial in ensuring the safety and survival of you and your loved ones.

Long-Term Radiation Exposure: What You Need to Know

While the immediate dangers of short-term radiation exposure are alarming and require urgent action, the long-term effects of radiation exposure can be equally concerning, albeit more insidious. Understanding these long-term risks is crucial for planning not just immediate survival strategies but also for ensuring the well-being of you and your loved ones in the months, years, and even decades after a nuclear event. This chapter explores the nature of long-term radiation exposure, its health implications, and the strategies you need to implement for sustained protection.

Long-term radiation exposure occurs when individuals or populations are subjected to lower levels of radiation over extended periods. Unlike acute exposure, which involves a high dose of radiation in a short time, chronic exposure involves continuous or intermittent contact with radioactive materials that remain in the environment after the initial fallout. This exposure can happen through various pathways, including the inhalation of contaminated dust, ingestion of contaminated food and water, and prolonged proximity to irradiated surfaces or soil.

One of the most significant concerns with long-term radiation exposure is the increased risk of cancer. Radiation is a potent carcinogen, meaning it can induce changes in cellular DNA that lead to the development of cancer. The types of cancer most commonly associated with radiation exposure include leukemia, thyroid cancer, breast cancer, lung cancer, and bone cancer. The risk of developing these cancers depends on several factors, including the level of exposure, the duration of exposure, and the specific radioactive isotopes involved.

For instance, iodine-131, a common by-product of nuclear explosions, tends to accumulate in the thyroid gland. Prolonged exposure to iodine-131 can lead to thyroid cancer, particularly in children and adolescents, whose thyroid glands are more sensitive to radiation. Similarly, strontium-90, another hazardous isotope, behaves like calcium and can become incorporated into bones, increasing the risk of bone cancer and leukemia over time.

The latency period between radiation exposure and the onset of cancer can vary widely. Some cancers, such as leukemia, may develop within a few years of exposure, while others, like solid tumors, may not appear until decades later. This long latency period makes it difficult to immediately assess the full impact of radiation exposure, underscoring the importance of ongoing monitoring and health surveillance for those exposed to fallout.

In addition to cancer, long-term radiation exposure can lead to a host of other chronic health issues. Cardiovascular diseases, including heart disease and stroke, have been linked to prolonged radiation exposure. Studies of atomic bomb survivors and those exposed to radiation in other contexts have shown that even moderate levels of radiation can increase the risk of heart-related conditions, likely due to the damage radiation inflicts on blood vessels and heart tissues.

Radiation can also have profound effects on the immune system, leading to a weakened ability to fight infections. This is particularly concerning in a post-nuclear environment, where medical resources may be limited, and the risk of secondary infections is high. The suppression of the immune system can make even minor illnesses more dangerous and can complicate recovery from injuries or other health conditions.

Another long-term concern is the potential for genetic damage. Radiation can cause mutations in the DNA of reproductive cells, leading to genetic defects that can be passed on to future generations. This risk is particularly relevant for individuals exposed during childhood or young adulthood, as these are the stages of life when reproductive cells are most active. The potential for genetic mutations raises ethical and practical concerns for families planning to have children after a nuclear event, as the effects of radiation may not be fully understood for years.

Environmental contamination is another critical aspect of long-term radiation exposure. Radioactive materials can persist in the environment for decades, contaminating soil, water, and air. This contamination can affect agriculture, leading to the production of food that is unsafe to eat due to the accumulation of radionuclides in plants and animals. Cesium-137, for example, can remain in the environment for decades, contaminating crops and livestock and posing a long-term threat to food security.

Water sources can also be contaminated by fallout, making it challenging to find safe drinking water. Radioactive particles can enter rivers, lakes, and groundwater, leading to widespread contamination that can affect entire regions. Long-term exposure to contaminated water can contribute to the cumulative radiation dose received by a population, increasing the risk of chronic health effects.

Given these risks, it's essential to implement strategies to minimize long-term radiation exposure.

One of the most effective ways to reduce exposure is through decontamination. This involves removing or neutralizing radioactive materials from surfaces, soil, and water. Decontamination can be as simple as washing contaminated clothing or skin, or as complex as removing topsoil from affected areas or treating contaminated water supplies. Regular decontamination efforts are crucial in a post-fallout environment to reduce the ongoing exposure risk.

Monitoring radiation levels over time is another critical strategy. Using dosimeters, Geiger counters, and other radiation detection devices, you can assess the safety of your environment and make informed decisions about where to live, work, and grow food. Continued monitoring is especially important because radiation levels can fluctuate as radioactive materials decay and as environmental factors like rain or wind redistribute fallout particles.

Long-term health surveillance is also vital for those who have been exposed to radiation. Regular medical check-ups, cancer screenings, and blood tests can help detect early signs of radiation-induced health issues, allowing for timely intervention. For those in high-risk groups, such as children and pregnant women, more frequent monitoring may be necessary to catch potential problems early.

In terms of sheltering, long-term strategies involve more than just immediate protection from fallout. Ensuring that your shelter is equipped for extended stays, with adequate food, water, medical supplies, and radiation shielding, is essential. Over time, you may also need to consider the location of your shelter, as areas with higher levels of contamination may require relocation to reduce exposure.

Education and community planning are also key components of long-term survival. By educating yourself and others about the risks of long-term radiation exposure and the importance of ongoing protective measures, you can contribute to a safer, more informed community. Working together, communities can develop plans for agriculture, water purification, and medical care that address the challenges of living in a contaminated environment.

In conclusion, while the immediate aftermath of a nuclear event demands quick action to protect against acute radiation exposure, the long-term risks require sustained vigilance and planning. Understanding the health implications of chronic radiation exposure, the environmental challenges it presents, and the strategies for mitigation is essential for ensuring the safety and well-being of you and your family in the years to come. By taking these long-term risks seriously and preparing accordingly, you can increase your chances of not just surviving, but thriving in a post-nuclear world.

The Spread of Fallout: How It Travels and Settles

When a nuclear explosion occurs, the immediate devastation is only the beginning of the dangers that follow. One of the most significant threats comes from nuclear fallout, the radioactive particles that are lifted into the atmosphere and eventually settle back to Earth. Understanding how fallout spreads and where it is likely to settle is crucial for planning effective protection strategies and ensuring long-term safety. This chapter explores the mechanisms that drive the spread of fallout, how it travels through the environment, and the factors that influence where it eventually settles.

The journey of fallout begins at the moment of the nuclear explosion. When a nuclear device detonates, it releases an enormous amount of energy in the form of heat, light, and radiation. This energy vaporizes the materials in the immediate vicinity of the explosion, including soil, buildings, and anything else in the blast zone. The resulting fireball rises rapidly into the atmosphere, carrying with it vast quantities of vaporized radioactive materials. As the fireball ascends, it cools and condenses, forming a mushroom cloud that contains a mix of radioactive particles and debris.

The height and shape of the mushroom cloud play a critical role in determining how far and wide the fallout will spread. In a ground-level explosion, the cloud tends to be lower and contains more debris from the ground, which results in heavier fallout that tends to settle closer to the explosion site. In contrast, an airburst—where the explosion occurs at a higher altitude—produces a higher mushroom cloud with lighter particles that can be carried much further by the winds before they settle.

Once the radioactive particles are in the atmosphere, they are at the mercy of the wind. Wind patterns are the primary driver of fallout distribution, and they can carry radioactive materials over vast distances, potentially affecting areas far removed from the initial explosion. The speed and direction of the wind at various altitudes influence where the fallout will travel. For instance, strong winds at high altitudes can carry fallout across continents, while lower-altitude winds might distribute particles over a more localized area.

The fallout particles eventually begin to settle back to the ground, a process influenced by several factors, including the size and weight of the particles, weather conditions, and the terrain over which they are traveling. Larger and heavier particles tend to fall back to Earth more quickly and close to the explosion site. These particles are typically composed of debris from the ground and are more likely to contain highly radioactive isotopes, making them particularly dangerous in the short term.

Smaller and lighter particles, on the other hand, can remain suspended in the atmosphere for longer periods and be carried further away by the wind. These particles can eventually settle in areas far from the blast site, potentially contaminating regions that were not directly affected by the explosion. This long-distance spread of fallout means that even areas hundreds or thousands of miles away from the explosion can experience radioactive contamination.

Weather conditions at the time of the explosion also significantly impact the spread of fallout. Rain or snow can cause fallout particles to be washed out of the atmosphere more quickly, a process known as "rainout." While this might seem like a good thing, rainout can concentrate fallout in specific areas, leading to "hot spots" of intense radiation that can be more dangerous than a more evenly distributed fallout pattern. In contrast, dry conditions can allow fallout to remain airborne for longer, spreading it over a wider area.

The terrain over which the fallout travels can also influence where it settles. Mountain ranges, valleys, and other geographical features can act as barriers or funnels, directing the flow of wind and potentially concentrating fallout in certain areas. Urban environments with tall buildings can create complex wind patterns that affect how and where fallout particles settle, potentially leading to unexpected contamination zones within cities.

One of the most concerning aspects of fallout is its ability to contaminate essential resources, such as water and food supplies. Fallout particles that settle on rivers, lakes, or reservoirs can lead to the contamination of drinking water. Similarly, particles that settle on agricultural land can contaminate crops and livestock, making food unsafe to eat. This contamination can persist for long periods, especially in areas where radionuclides with long half-lives, like cesium-137 or strontium-90, have settled.

Understanding the potential spread of fallout is crucial for planning protective measures, such as the location of fallout shelters and the strategies for decontamination and long-term survival. Ideally, fallout shelters should be located in areas less likely to be heavily contaminated by fallout, taking into account prevailing wind patterns, terrain features, and the potential for rainout events. Additionally, knowing how fallout spreads can help in planning evacuation routes and timing, ensuring that people move away from areas that are likely to become contaminated.

In the days and weeks following a nuclear explosion, monitoring the spread of fallout is vital for assessing the safety of different areas. Government agencies and emergency services typically track radiation levels and provide information on which areas are safe and which are not. However, having your own radiation detection equipment, such as a Geiger counter or dosimeter, can provide real-time information and help you make informed decisions about where to go and how long to stay sheltered.

Decontamination efforts are also influenced by the spread of fallout. Areas where fallout has settled heavily will require more intensive decontamination, which might involve removing topsoil, cleaning surfaces, or treating water supplies. In contrast, areas with lighter fallout might only need minimal intervention. Understanding how fallout settles can also help in predicting which areas will recover more quickly and which will remain hazardous for longer periods.

Another important consideration is the re-suspension of fallout particles. Wind, rain, or human activity can stir up settled fallout, causing it to become airborne again and potentially leading to additional exposure. This re-suspension can be particularly problematic in dry or windy conditions, where particles are easily disturbed. Ongoing monitoring and dust control measures, such as wetting down surfaces, can help reduce the risk of re-suspension.

In conclusion, the spread and settlement of fallout are complex processes influenced by a range of factors, including the nature of the explosion, wind patterns, weather conditions, and terrain. Understanding these processes is essential for planning effective protection strategies, from the location of fallout shelters to the management of contaminated resources.

Protecting Your Home: The Importance of Shielding

In the aftermath of a nuclear explosion, your home can either be a sanctuary of safety or a place of peril, depending on how well it is protected from radioactive fallout. Shielding is one of the most critical aspects of fallout protection, as it directly impacts your ability to survive the initial fallout and endure the long-term radiation that can persist for weeks, months, or even years. This chapter will delve into the importance of shielding, exploring the principles behind it, the materials that offer the best protection, and practical steps you can take to fortify your home.

At its core, shielding is about creating a barrier between you and the harmful radiation emitted by fallout particles. Radiation comes in different forms—alpha particles, beta particles, and gamma rays—with gamma rays being the most penetrating and thus the most dangerous. Effective shielding must be able to block or at least reduce the intensity of these gamma rays to protect the occupants of a shelter.

The effectiveness of a material as a shield is measured by its ability to reduce the intensity of radiation, a concept known as attenuation. The thicker and denser a material is, the better it will attenuate, or weaken, the radiation passing through it. This is why materials like lead, concrete, and even earth are commonly used in the construction of fallout shelters—they have the density and mass necessary to absorb or deflect gamma rays, thereby reducing the radiation dose that reaches you.

When considering the shielding for your home, one of the first steps is to assess the materials already in place. Many homes are constructed with materials like wood, drywall, and glass, which offer minimal protection against gamma rays. However, these materials can still play a role in shielding when used in combination with more effective barriers. For instance, adding layers of protection—such as placing heavy furniture against walls, covering windows with thick blankets, or even piling earth against the exterior walls—can enhance the overall shielding effect.

Concrete is one of the most effective and accessible materials for shielding. A few inches of concrete can significantly reduce the intensity of gamma radiation. In practical terms, this means that a basement with thick concrete walls and a concrete ceiling can serve as an effective fallout shelter, especially if the entrance is well-sealed and the space is stocked with supplies. If your home has a basement, reinforcing it with additional concrete layers, such as building interior walls or adding a concrete floor, can further improve its protective capabilities.

Lead is another highly effective shielding material, but it is often less practical for large-scale use due to its weight and cost. However, lead can be used in smaller, critical areas where maximum protection is needed, such as a designated safe room within your home. Lead sheeting or lead bricks can be installed in walls or around key areas to create a concentrated zone of protection.

Earth and soil also offer excellent shielding properties, and they are readily available in most locations. In fact, many traditional fallout shelters are built underground, using the earth as the primary shield. Even above-ground homes can benefit from the use of earth as a shielding material.

For example, you can create berms—mounds of earth—against the exterior walls of your home to add an extra layer of protection. This method is particularly effective for homes that do not have basements, as the berms can help reduce the radiation that penetrates the lower levels of the house.

Bricks, especially dense bricks like those used in older homes, also provide good shielding. If your home is made of brick, you already have a level of protection that is superior to wood or vinyl siding. Enhancing this by adding additional layers, such as insulating with concrete or lining the interior walls with additional brick or stone, can further increase your home's resistance to radiation.

While shielding materials like concrete, lead, and earth are effective, the key to maximizing their protective value lies in how they are used. The principle of "mass thickness" comes into play here, which essentially means that the more material you have between you and the source of radiation, the better protected you are. For example, if you are constructing a fallout shelter within your home, aim for walls that are as thick as possible, using the densest materials available to you.

Windows and doors are often the weakest points in a home's shielding. Glass provides almost no protection against gamma rays, so windows should be covered with heavy materials—such as metal sheets, thick wood, or even lead blankets—during a fallout event. Doors should be reinforced, sealed, and ideally backed with additional layers of protective material. In some cases, it might be necessary to block off certain rooms entirely, creating a smaller, more secure space within the home where your family can shelter during the most dangerous periods of fallout.

Ventilation is another important consideration. While it's crucial to have access to fresh air, any openings in your home—such as windows, doors, and vents—can allow radioactive particles to enter. During a fallout event, these openings should be sealed as much as possible, using duct tape, plastic sheeting, or even wet towels to block any gaps. In more advanced shelters, air filtration systems equipped with HEPA filters or activated charcoal can help remove radioactive particles from the air, allowing for safer ventilation.

It's also important to consider the duration of your stay in a sheltered environment. The initial fallout after a nuclear explosion is the most dangerous, with radiation levels peaking within the first 24 to 48 hours. However, radioactive decay will cause these levels to decrease over time, which means that the intensity of the radiation you are exposed to will gradually diminish. This is why it's often recommended to stay in your shelter for at least 48 hours before considering venturing outside, and longer if possible.

In summary, shielding your home from nuclear fallout is a critical step in ensuring your safety and the safety of your loved ones. By understanding the principles of radiation attenuation and using the right materials, you can significantly reduce the radiation levels inside your home, providing a safer environment during a fallout event. Whether you are enhancing an existing structure or planning a dedicated fallout shelter, the importance of effective shielding cannot be overstated. With the right preparations, your home can become a bastion of safety in the face of one of the most severe threats imaginable.

Building Materials: Plastic and Its Protective Qualities

When thinking about materials for constructing a fallout shelter or fortifying your home against radioactive contamination, plastic might not be the first material that comes to mind. However, while plastic isn't as effective at blocking radiation as denser materials like concrete or lead, it does have specific qualities that make it valuable in the broader context of fallout protection. This chapter will explore the protective qualities of plastic, how it can be used effectively in conjunction with other materials, and its role in creating a safer environment during a nuclear fallout event.

To understand the role of plastic in radiation protection, it's essential to recognize what it can and cannot do. Plastic, in its various forms, is not dense enough to provide substantial shielding against the most penetrating types of radiation, such as gamma rays. Gamma rays require thick layers of dense materials to attenuate effectively. However, plastic does offer several protective benefits that can be crucial in a fallout scenario.

One of the primary benefits of plastic is its ability to act as a barrier against contamination. Fallout consists of radioactive particles that can settle on surfaces, infiltrate spaces, and contaminate food, water, and air. Plastic sheeting, tarps, and covers can be used to create barriers that prevent these particles from coming into direct contact with you or your belongings. For example, plastic sheeting can be used to seal windows, doors, and vents, preventing radioactive dust from entering your living space. This containment of radioactive particles is a critical first line of defense, especially in the immediate aftermath of a fallout event when radiation levels are highest.

Plastic is also highly versatile and easy to work with, making it an excellent material for temporary structures or quick fixes. In a fallout situation, you might not have the time or resources to construct complex barriers or decontamination zones. Plastic sheeting or tarps can be quickly deployed to cover surfaces, protect food supplies, and create makeshift shelters or decontamination areas. This versatility can be a lifesaver in situations where you need to act fast to protect yourself from radiation exposure.

Another significant advantage of plastic is its resistance to moisture and chemicals. In a fallout scenario, water contamination is a major concern, and plastic can help protect against this. For instance, plastic liners can be used to safeguard water storage containers, ensuring that they remain free from radioactive particles. Similarly, plastic coverings can protect food supplies and other essential items from contamination. This moisture resistance also means that plastic can be used to create effective barriers against radioactive rain or snow, which can occur if fallout particles are carried into precipitation.

While plastic alone is not sufficient to block radiation, it can be combined with other materials to enhance its protective qualities. For example, plastic sheeting can be used in conjunction with earth, sandbags, or concrete to create layered barriers that offer better protection against radiation. In this setup, the plastic serves as a containment layer, preventing radioactive particles from penetrating deeper into the shelter or living area, while the denser materials provide the necessary shielding against gamma rays.

Plastic also plays a critical role in decontamination efforts. After a fallout event, one of the priorities is to remove or neutralize radioactive particles that have settled on surfaces. Plastic sheeting can make this process easier by acting as a disposable surface covering. By laying plastic over floors, walls, and furniture, you create a layer that can be easily removed and disposed of, taking the radioactive particles with it. This simplifies the decontamination process and reduces the risk of spreading contamination to other areas.

In addition to its practical applications, plastic is relatively inexpensive and widely available, making it accessible to most people preparing for a fallout event. Rolls of plastic sheeting, garbage bags, and tarps are common items that can be purchased in bulk and stored easily. This affordability means that you can stockpile plastic materials without a significant financial burden, ensuring that you have enough on hand to cover windows, doors, and other vulnerable areas in your home.

However, it's important to recognize the limitations of plastic in a nuclear fallout scenario. While plastic can prevent radioactive particles from entering your living space and contaminating essential supplies, it cannot shield you from the radiation emitted by those particles. Therefore, plastic should be viewed as a supplementary material in your fallout protection strategy, one that is used in conjunction with more robust shielding materials.

When using plastic in your fallout preparations, it's also essential to consider how to secure it properly. Plastic sheeting must be tightly sealed to prevent any gaps where particles could enter. This can be done using duct tape, staples, or other fasteners, depending on the situation. For windows and doors, it's advisable to overlap the plastic sheeting and double-tape the edges to ensure a tight seal. If possible, creating airlocks or double layers of plastic barriers can provide an additional level of protection, reducing the likelihood of contamination.

In summary, while plastic may not be the ultimate shielding material for radiation, it offers unique and valuable qualities that make it an essential component of a comprehensive fallout protection plan. Its ability to act as a barrier against contamination, its versatility, and its resistance to moisture and chemicals make plastic an indispensable material for protecting your home and belongings during a fallout event. By understanding the strengths and limitations of plastic, and by combining it with other shielding materials, you can create a safer environment for yourself and your family in the face of nuclear fallout.

How Glass Holds Up Against Radiation

Glass, a ubiquitous material in modern homes and buildings, is often overlooked when considering protection against radiation. However, in the context of nuclear fallout, understanding how glass interacts with radiation is crucial for assessing the vulnerabilities of your home and planning effective protective measures. This chapter explores the properties of glass, its limitations as a shielding material, and how it can be reinforced or used strategically in your fallout protection plan.

To begin with, it's important to recognize that glass, like plastic, is not a dense material. This lack of density makes it a poor barrier against gamma rays, the most penetrating type of radiation produced by nuclear fallout. Gamma rays can pass through glass with minimal attenuation, meaning that glass alone provides little to no protection against the harmful effects of radiation. This is a significant concern for homes with large windows or glass doors, as these areas could be points of vulnerability during a fallout event.

Despite its limitations, glass does have some properties that are worth considering in the broader context of fallout protection. First, glass is impervious to the infiltration of radioactive particles. Fallout particles cannot penetrate glass, which means that glass surfaces can help keep radioactive dust and debris out of your home when windows and doors are sealed properly. This is particularly important during the initial stages of fallout when radioactive particles are most concentrated in the environment.

However, the challenge with glass lies in its transparency and fragility. The very properties that make glass desirable for letting in light and providing views also make it a weak point in radiation protection. Transparent glass allows radiation to pass through unimpeded, and in the event of a nuclear explosion, the intense blast wave can shatter glass windows, potentially allowing radioactive particles to enter your home.

Given these vulnerabilities, one of the primary considerations for using glass in a fallout protection scenario is reinforcement. There are several strategies to reinforce glass and mitigate the risks it poses during a nuclear event.

One approach is to apply protective films to windows. These films are typically made of polyester and are designed to hold glass fragments together if the window is shattered by a blast wave. While these films do not stop radiation, they can prevent glass from breaking into dangerous shards and can help keep the radioactive particles outside. Additionally, certain types of protective films are designed to reduce the amount of ultraviolet (UV) and infrared (IR) radiation that passes through the glass, though their effectiveness against gamma rays is minimal.

Another strategy is to cover windows with additional layers of protective material. During a fallout event, windows can be sealed with plastic sheeting, heavy curtains, or even plywood to block the entry of radioactive particles and reduce the amount of radiation that enters the home. The goal here is not to rely on the glass itself for protection but to use it as a base layer that is reinforced with other materials to enhance shielding and containment.

For homes with large windows or glass doors, it may be worth considering more permanent solutions, such as installing shutters or storm panels. These can be closed and secured during a fallout event to provide a more substantial barrier against radiation and blast effects. In addition, storm panels made of materials like steel or aluminum can offer some degree of radiation shielding, although they are primarily designed to protect against the physical impact of a blast.

In more extreme cases, especially in high-risk areas, replacing standard glass with leaded glass is an option. Leaded glass, often used in medical and industrial applications, contains a higher density of lead oxide, which gives it the ability to attenuate gamma radiation more effectively than regular glass. While leaded glass is significantly heavier and more expensive, it can provide better protection in scenarios where high levels of radiation are expected. However, the use of leaded glass in residential settings is typically limited due to cost and practicality.

Another aspect to consider is the strategic placement of glass in your home or fallout shelter. If possible, areas that are intended for sheltering should have minimal exposure to exterior windows or glass doors. Interior rooms with fewer or no windows are preferable for sheltering, as they naturally provide better protection against radiation. If your home's layout necessitates sheltering in a room with windows, consider boarding up the windows or creating an additional barrier inside the room using furniture, bookshelves, or other dense materials.

Ventilation is another factor to consider when dealing with glass in a fallout scenario. While sealing windows is crucial to prevent the entry of radioactive particles, you also need to ensure that your shelter remains ventilated. In some cases, special vents or air filtration systems can be installed to allow for airflow without compromising the integrity of the sealed environment. These systems often use HEPA filters to capture fine particles, including radioactive dust, while allowing air to circulate.

In summary, while glass is not a suitable material for shielding against radiation, it still plays a role in fallout protection by acting as a barrier against radioactive particles. Its primary vulnerabilities—transparency to gamma rays and susceptibility to shattering—can be mitigated through reinforcement with protective films, coverings, or shutters. In high-risk situations, leaded glass may offer enhanced protection, though it is not a common choice for residential settings. By understanding the limitations and possibilities of glass in the context of radiation protection, you can make informed decisions about how to fortify your home and ensure a safer environment during a nuclear fallout event.

The Role of Concrete in Fallout Shelters

Concrete is one of the most effective and widely used materials in the construction of fallout shelters, primarily because of its excellent shielding properties against radiation. In the context of nuclear fallout, where the primary concern is protecting against gamma rays, concrete offers a robust defense due to its density and mass. This chapter explores the role of concrete in fallout shelters, examining why it is so effective, how it should be used, and the practical considerations involved in incorporating concrete into your shelter design.

To understand why concrete is so valuable in fallout protection, it's essential to revisit the basic principles of radiation shielding. Gamma rays, the most penetrating type of radiation produced by nuclear fallout, require dense materials to attenuate their energy. The denser the material, the more effective it is at absorbing or deflecting gamma rays, reducing the radiation dose that reaches people sheltered behind it. Concrete, with its high density and availability, fits this requirement perfectly.

One of the key advantages of concrete is its mass thickness, a concept that refers to the amount of material mass between you and the source of radiation. Mass thickness is crucial because it determines the extent to which radiation is attenuated. In practical terms, a thick wall of concrete can reduce the intensity of gamma rays to a fraction of their original strength, making it one of the most reliable materials for fallout shelters.

The thickness of concrete required to provide adequate radiation shielding depends on the level of radiation expected. A general rule of thumb is that every 2 inches of concrete reduces gamma radiation by approximately 50%. This means that a concrete wall that is 12 inches thick can reduce gamma radiation to less than 5% of its original intensity. For a fallout shelter, a concrete wall of at least 12 to 24 inches is typically recommended, depending on the expected exposure and available resources.

Concrete's effectiveness as a shielding material also comes from its ability to be reinforced and combined with other materials. Reinforced concrete, which includes steel rebar, offers additional strength and durability, making it ideal for constructing walls, ceilings, and floors that must withstand not only radiation but also potential blast effects. The combination of concrete's density with the tensile strength of steel creates a structure that can endure significant physical stress while providing excellent radiation protection.

In the design and construction of a fallout shelter, concrete is often used in multiple ways. The most common application is in the walls and ceiling of the shelter. A concrete basement, for example, can be an excellent starting point for a fallout shelter because the existing concrete foundation already provides a significant level of protection. By reinforcing the basement walls with additional layers of concrete or by adding a concrete ceiling, you can enhance the shelter's ability to protect against both radiation and physical impacts.

Another practical use of concrete in fallout shelters is in the construction of shielding barriers around vulnerable areas, such as entrances, ventilation systems, and escape hatches. These areas can be weak points in the shelter's overall protection if not properly fortified. By building concrete barriers or blast shields, you can ensure that these entry points do not become liabilities during a fallout event. These barriers can be designed to be movable or permanent, depending on the shelter's design and intended use.

The floor of the shelter is another critical area where concrete plays a role. A concrete floor not only provides a stable and durable base but also contributes to the overall shielding effect. In some designs, the floor may be thickened or reinforced with additional concrete to prevent radiation from penetrating from below, particularly if the shelter is built above ground or in an area with a high water table that could potentially introduce contamination.

Concrete is also used in the construction of water storage tanks, waste management systems, and other essential infrastructure within the shelter. Because concrete is non-porous and resistant to chemical reactions, it is ideal for creating long-lasting, secure storage for potable water and for managing waste products that must be contained within the shelter for extended periods.

In addition to its shielding properties, concrete offers several practical benefits that make it a preferred material for fallout shelters. It is fire-resistant, which is crucial in the aftermath of a nuclear explosion where fires may be a secondary threat. Concrete is also resistant to moisture and pests, ensuring that the shelter remains dry and habitable over the long term. Furthermore, concrete can be cast into almost any shape, allowing for flexibility in shelter design to accommodate specific needs and constraints.

However, while concrete is an excellent material for fallout protection, there are practical considerations and limitations to keep in mind. The construction of thick concrete walls and ceilings requires significant planning and resources. The weight of concrete can impose structural challenges, particularly in retrofitting existing buildings or adding additional layers to an existing structure. In some cases, the foundation may need to be reinforced to support the additional load, which can add to the cost and complexity of the project.

Additionally, the curing process of concrete—where it hardens and gains strength—takes time and must be carefully managed to avoid cracks or weaknesses in the structure. Improperly cured concrete can compromise the integrity of the shelter, reducing its effectiveness in protecting against radiation and physical impacts. It's important to work with experienced contractors or engineers when building or retrofitting a concrete fallout shelter to ensure that the construction meets the necessary safety standards.

Cost is another factor to consider. While concrete is relatively affordable compared to other shielding materials like lead, the amount required for a fully effective fallout shelter can be substantial. Budgeting for the construction of a concrete shelter should include not just the cost of materials but also labor permits, and potential reinforcement of existing structures.

In conclusion, concrete plays a vital role in the construction of effective fallout shelters due to its excellent shielding properties against gamma radiation, its strength, and its versatility. Whether you are building a dedicated fallout shelter from scratch or retrofitting an existing structure, concrete should be a central component of your design. By understanding the principles behind concrete's effectiveness, as well as the practical considerations involved in its use, you can create a shelter that provides robust protection in the event of nuclear fallout, ensuring the safety and survival of you and your family.

Wood and Plaster: Natural Defenses

Wood and plaster are common building materials that are found in many homes and structures. While they are not as effective as concrete or lead in protecting against radiation, they do offer certain protective qualities that can contribute to the overall defense of your home or fallout shelter. This chapter explores the role of wood and plaster in fallout protection, how they can be utilized effectively, and their limitations in shielding against radiation.

Understanding the Properties of Wood and Plaster

To begin with, it's important to recognize that both wood and plaster are relatively low-density materials. Radiation shielding is largely a function of material density; the denser the material, the better it is at blocking or attenuating radiation, particularly gamma rays. Wood and plaster, being less dense than materials like concrete or metal, provide less attenuation of gamma rays. This means that, on their own, wood and plaster are not sufficient to offer significant protection against high levels of radiation.

However, both materials do offer some degree of protection, particularly when used in combination with other, denser materials. Additionally, wood and plaster can help protect against other fallout-related hazards, such as airborne radioactive particles, and can serve as supplementary layers in a more comprehensive shielding strategy.

The Role of Wood in Fallout Protection

Wood is a versatile and widely available material that is often used in the construction of walls, floors, and roofs. While it does not offer significant protection against gamma radiation, wood can still play a role in fallout protection through its use in construction and as part of layered shielding strategies.

One of the primary advantages of wood is its ability to act as a barrier against radioactive particles. In a fallout scenario, radioactive dust and debris can settle on surfaces, and wood can help prevent these particles from penetrating further into the structure. For example, wooden walls and floors can help contain radioactive particles, especially if they are properly sealed and maintained. This containment is particularly important in areas where the fallout may linger or accumulate, such as in attics, basements, or other enclosed spaces.

Wood can also be used in combination with other materials to enhance overall shielding. For example, wooden walls can be reinforced with additional layers of material, such as plywood, drywall, or even metal sheeting, to create a more substantial barrier. When used in conjunction with denser materials like concrete or sandbags, wood can contribute to a layered shielding strategy that reduces radiation exposure.

Another practical use of wood in a fallout shelter is in the construction of interior structures, such as furniture, storage areas, and partition walls. These structures can add mass to the shelter, providing some additional shielding, and can also help organize the space efficiently. Additionally, wood is relatively easy to work with, making it a convenient material for making quick modifications or repairs to the shelter.

The Role of Plaster in Fallout Protection

Plaster, like wood, is a common building material, particularly in older homes where it is often used as a wall covering. Plaster is typically made from a mixture of lime or gypsum and water, applied in layers over wooden laths or metal mesh. While plaster itself is not particularly dense, it does offer some level of protection against radiation, especially when applied in thick layers.

One of the key benefits of plaster is its ability to create a smooth, sealed surface that can help prevent the infiltration of radioactive particles. A well-applied layer of plaster can act as a barrier, sealing gaps and cracks in walls and ceilings where radioactive dust might otherwise enter. Additionally, because plaster is typically applied in multiple layers, it can add some mass to the walls, enhancing the overall attenuation of radiation.

In some cases, plaster can be reinforced with other materials to improve its protective qualities. For example, applying a layer of plaster over a dense substrate, such as brick or concrete, can increase the wall's overall effectiveness in blocking radiation. This layered approach can be particularly useful in retrofitting existing structures for fallout protection.

Plaster can also be used to seal and finish other protective materials, such as sandbags or concrete blocks, providing a more aesthetically pleasing and smooth surface while still contributing to the overall shielding effect. This can be particularly beneficial in shelters where comfort and habitability are important for long-term stays.

Limitations of Wood and Plaster

While wood and plaster offer some protective benefits, it's important to understand their limitations. Neither material is dense enough to provide significant attenuation of gamma rays, the most penetrating form of radiation associated with nuclear fallout. Therefore, they should not be relied upon as primary shielding materials in a fallout shelter.

However, when used as part of a layered defense strategy, wood and plaster can contribute to the overall protection of a shelter by adding mass, sealing gaps, and preventing the infiltration of radioactive particles. They are most effective when used in conjunction with other materials, such as concrete, lead, or earth, which provide the primary radiation shielding.

Another limitation of wood is its susceptibility to fire. In the aftermath of a nuclear explosion, fires may be a secondary threat, and wood can be a combustible material. It's important to consider fire safety when using wood in the construction of a fallout shelter, particularly in areas where there may be an increased risk of fire.

Plaster, while non-combustible, can be prone to cracking over time, particularly in areas subject to moisture or structural movement. Cracks in plaster can create pathways for radioactive particles to enter the shelter, reducing its effectiveness as a barrier. Regular maintenance and inspection of plaster surfaces are essential to ensure that they remain intact and effective.

Practical Applications in Shelter Design

When designing a fallout shelter or retrofitting an existing structure, wood and plaster can be effectively utilized in various ways. For example, a wooden structure can be reinforced with layers of plaster to create a more substantial barrier against radioactive particles. Similarly, wooden furniture and storage units can be strategically placed to add mass and provide additional shielding within the shelter.

Plaster can be used to seal and finish walls, creating a smooth, continuous surface that prevents the infiltration of dust and debris. In areas where additional shielding is required, plaster can be applied over denser materials to create a layered defense that enhances radiation protection. While wood and plaster are not the most effective materials for shielding against radiation, they do offer important protective qualities that can be utilized in a fallout shelter.

Bricks and Mortar: A Solid Foundation for Safety

Bricks and mortar have been used in construction for centuries, valued for their durability, strength, and availability. In the context of nuclear fallout protection, these materials play a crucial role in providing a solid foundation for safety. While not as dense as concrete, bricks and mortar offer significant protective qualities that can be leveraged in building or reinforcing a fallout shelter. This chapter will explore the benefits of using bricks and mortar in fallout protection, their effectiveness as shielding materials, and practical considerations for incorporating them into your shelter design.

The Protective Qualities of Bricks

Bricks are primarily made from clay or shale that has been fired at high temperatures, resulting in a hard, dense material. This density gives bricks a level of effectiveness in attenuating radiation, particularly gamma rays, which are the most penetrating type of radiation produced by nuclear fallout. While bricks do not offer the same level of protection as thicker concrete or lead, they are still a valuable material for constructing barriers that can reduce radiation exposure.

One of the key advantages of bricks is their mass. The more mass a material has, the better it is at absorbing or deflecting gamma rays, thus reducing the amount of radiation that passes through. A single layer of brick will not block all radiation, but multiple layers or a thick wall made from bricks can significantly attenuate the radiation, making it safer for those inside.

For example, a standard brick wall that is 12 inches thick can reduce the intensity of gamma radiation to about 50% of its original level. While this alone might not be sufficient for a fallout shelter in a high-radiation environment, when combined with other materials or additional layers, brick walls can contribute to a more comprehensive shielding strategy.

Mortar: The Bond That Holds It All Together

Mortar, the material used to bond bricks together, is typically composed of a mixture of sand, cement, and water. While mortar itself does not contribute significantly to radiation shielding, it is essential for the structural integrity of a brick wall. A well-constructed brick wall with strong mortar joints will be more effective at blocking radiation because it minimizes gaps and weak points where radiation could penetrate.

Mortar also adds some mass to the wall, albeit to a lesser extent than the bricks themselves. In fallout shelter construction, ensuring that mortar joints are well-filled and properly cured is crucial for maximizing the wall's protective capabilities. Poorly applied mortar can lead to cracks or gaps that reduce the overall effectiveness of the brick wall as a shield.

Combining Bricks and Mortar with Other Materials

One of the strengths of bricks and mortar in fallout protection is their versatility. They can be combined with other materials to enhance their protective qualities. For example, a brick wall can be backed with concrete or covered with additional layers of protective materials, such as earth or sandbags, to increase the overall shielding effect.

In some cases, bricks can be used to construct an outer shell, while the interior of the wall is filled with a denser material like concrete or gravel. This combination leverages the structural strength of bricks and mortar while benefiting from the superior shielding properties of denser materials. This technique is particularly useful in situations where budget or availability of materials is a concern, allowing you to create a robust shelter with what you have on hand.

Practical Considerations for Using Bricks and Mortar

When planning to use bricks and mortar in your fallout shelter, there are several practical considerations to keep in mind:

Thickness: The thickness of the brick walls is a critical factor in determining the level of radiation protection. For optimal protection, aim for walls that are at least 12 inches thick, and consider adding additional layers or combining the brick with other materials to enhance shielding.

Quality of Materials: Not all bricks are created equal. Dense, well-fired bricks provide better radiation protection than lighter, more porous bricks. Similarly, the quality of the mortar used can impact the effectiveness of the wall. Using high-quality mortar and ensuring that it is applied correctly will improve the overall durability and protective capabilities of the wall.

Construction Techniques: The method of construction also plays a significant role. Bricks should be laid tightly together with fully filled mortar joints to prevent any gaps that could allow radiation to penetrate. It's also important to ensure that the walls are straight, level, and structurally sound, as any weaknesses in the construction could compromise the shelter's effectiveness.

Reinforcement: Depending on the specific design of your shelter, it may be necessary to reinforce the brick walls with additional support, such as steel rebar or concrete pillars. This reinforcement can help the walls withstand not only the weight of the structure but also any potential blast effects from a nearby nuclear explosion.

Maintenance: Over time, bricks and mortar can degrade, especially if exposed to moisture or extreme temperatures. Regular maintenance, such as inspecting for cracks or signs of wear, and making necessary repairs, is essential to ensure that the walls continue to provide effective protection.

Advantages of Brick and Mortar Construction

In addition to their shielding capabilities, bricks and mortar offer several other advantages that make them a good choice for fallout shelter construction:

Durability: Brick and mortar structures are incredibly durable and can last for decades or even centuries with proper maintenance. This longevity makes them ideal for a fallout shelter that may need to remain functional for an extended period.

Fire Resistance: Both bricks and mortar are non-combustible materials, providing excellent fire resistance. This is particularly important in a fallout scenario where fires may be a secondary hazard.

Availability: Bricks and mortar are widely available in most regions, making them an accessible and affordable option for many people. Whether you're building a new shelter or retrofitting an existing structure, these materials are often easy to source.

Aesthetic Versatility: While aesthetics might not be a primary concern in a fallout shelter, bricks do offer some versatility in terms of appearance. A brick-lined shelter can be made to feel more like a liveable space, which can be important for long-term comfort if the shelter is to be occupied for an extended time.

Using Brick and Mortar in Existing Structures

If you're retrofitting an existing building to serve as a fallout shelter, brick and mortar can be valuable materials for reinforcing walls or creating new protective barriers. For example, you might add a brick veneer to an existing wooden or drywall structure, thereby increasing its mass and improving its ability to attenuate radiation. Additionally, brick and mortar can be used to seal off windows or other vulnerable openings, further enhancing the shelter's overall protection.

Bricks and mortar form a solid foundation for fallout protection, offering a combination of durability, availability, and reasonable shielding capabilities. While they may not provide the same level of radiation attenuation as denser materials like concrete, they are nonetheless an effective component of a layered defense strategy.

Evaluating Modern Building Materials for Fallout Protection

In the rapidly evolving world of construction, modern building materials have become increasingly sophisticated, offering new possibilities for enhancing the safety and durability of structures. When it comes to fallout protection, evaluating these modern materials is essential to determine their effectiveness in shielding against radiation and providing long-term survivability in a post-nuclear environment. This chapter explores various contemporary building materials, assessing their strengths, limitations, and potential applications in the construction or retrofitting of fallout shelters.

The Importance of Density and Composition

As discussed in previous chapters, the effectiveness of a material in protecting against radiation largely depends on its density. Dense materials are better at attenuating gamma rays, the most penetrating form of radiation produced by nuclear fallout. Therefore, when evaluating modern building materials for fallout protection, density is a key factor to consider. However, other properties, such as durability, ease of installation, and resistance to environmental factors, also play significant roles in determining a material's suitability for a fallout shelter.

High-Performance Concrete

While traditional concrete has long been a staple in fallout shelter construction, advancements in concrete technology have led to the development of high-performance concrete (HPC). HPC is designed to offer superior strength, durability, and resistance to environmental stressors compared to conventional concrete. This type of concrete typically incorporates additional materials, such as silica fume, fly ash, or fibers, to enhance its properties.

Advantages:

Increased Strength: HPC can achieve compressive strengths much higher than traditional concrete, allowing for thinner walls that still provide excellent radiation shielding.

Durability: HPC is more resistant to cracking, chemical attacks, and water penetration, making it ideal for long-term use in harsh environments.

Density: The increased density of HPC improves its ability to attenuate gamma radiation, providing better protection in a fallout scenario.

Limitations:

Cost: HPC is generally more expensive than traditional concrete due to the specialized materials and techniques required for its production.

Complexity: The mix design and curing process for HPC are more complex, requiring skilled labor and precise control to achieve the desired properties.

Applications: HPC is an excellent choice for constructing critical components of a fallout shelter, such as walls, ceilings, and barriers, where maximum protection is required. It is particularly well-suited for environments where long-term durability and resistance to environmental degradation are priorities.

Ultra-High Performance Concrete (UHPC)

Taking concrete technology even further, Ultra-High Performance Concrete (UHPC) is a cutting-edge material known for its exceptional strength, durability, and resistance to environmental factors. UHPC typically contains a high percentage of fine materials, such as silica fume and high-range water reducers, which contribute to its superior performance.

Advantages:

Exceptional Strength: UHPC can achieve compressive strengths up to 200 MPa (29,000 psi), far surpassing traditional concrete and HPC.

Durability: UHPC is highly resistant to impact, abrasion, and chemical attacks, making it ideal for extreme environments.

Density: The high density of UHPC makes it one of the best materials for radiation shielding, offering superior protection in a thinner profile.

Limitations:

High Cost: UHPC is significantly more expensive than traditional concrete and HPC, making it less accessible for large-scale projects.

Specialized Handling: UHPC requires precise mixing, placing, and curing procedures, often necessitating the involvement of experts.

Applications: Due to its high cost and specialized nature, UHPC is best used in critical areas where maximum protection and durability are essential, such as in the construction of key structural elements or high-risk facilities. For a fallout shelter, UHPC could be used for reinforced walls, ceilings, or barriers in areas expected to experience the highest radiation levels or blast impacts.

Structural Insulated Panels (SIPs)

Structural Insulated Panels (SIPs) are a modern building material that consists of a rigid foam core sandwiched between two layers of structural board, typically oriented strand board (OSB). SIPs are known for their excellent thermal insulation properties and ease of construction.

Advantages:

Insulation: SIPs provide excellent thermal insulation, which can help maintain a stable temperature in a fallout shelter, especially in extreme weather conditions.

Ease of Installation: SIPs are prefabricated and can be quickly assembled on-site, reducing construction time and labor costs.

Strength: SIPs offer good structural strength, making them suitable for walls and roofs in shelter construction.

Limitations:

Radiation Shielding: The primary limitation of SIPs is their low density, which makes them ineffective as a primary radiation shield. Additional materials would be needed to provide adequate protection against gamma rays.

Fire Resistance: The foam core in SIPs is combustible, which could pose a fire risk in certain scenarios.

Applications: SIPs can be used in the construction of interior walls, partitions, or roofs where thermal insulation is a priority. However, they should be combined with denser materials, such as concrete or brick, to provide adequate radiation protection.

Autoclaved Aerated Concrete (AAC)

Autoclaved Aerated Concrete (AAC) is a lightweight, precast building material that offers good insulation properties and ease of installation. AAC is made by combining cement, lime, silica, and aluminium powder, which reacts to create air bubbles within the material, giving it a porous structure.

Advantages:

Lightweight: AAC is much lighter than traditional concrete, making it easier to handle and install.

Thermal Insulation: The porous structure of AAC provides excellent thermal insulation, helping to regulate the temperature within a shelter.

Fire Resistance: AAC is non-combustible and offers good fire resistance.

Limitations:

Radiation Shielding: Due to its porous structure, AAC has a lower density than traditional concrete, making it less effective at shielding against gamma rays. It would need to be used in combination with denser materials to provide adequate protection.

Structural Strength: While AAC is strong enough for many applications, it is not as strong as traditional concrete and may require additional reinforcement in high-load areas.

Applications: AAC can be used for non-load-bearing walls, partitions, or as an insulating layer within a composite wall system. It is best used in combination with other materials that provide the necessary radiation shielding.

Fiber-Reinforced Polymers (FRPs)

Fiber-Reinforced Polymers (FRPs) are composite materials made by embedding fibers (such as glass, carbon, or aramid) in a polymer matrix. FRPs are known for their high strength-to-weight ratio, corrosion resistance, and flexibility.

Advantages:

High Strength: FRPs offer excellent tensile strength, making them useful for reinforcing structures.

Corrosion Resistance: FRPs are resistant to corrosion, which is beneficial in environments exposed to moisture or chemicals.

Lightweight: FRPs are much lighter than traditional building materials, reducing the overall load on a structure.

Limitations:

Radiation Shielding: FRPs are not dense enough to provide significant radiation shielding and must be combined with other materials for effective fallout protection.

Cost: FRPs can be expensive, particularly when using high-performance fibers like carbon or aramid.

Applications: FRPs are best used as reinforcing elements within a fallout shelter, such as in the reinforcement of concrete walls, ceilings, or beams. They can also be used in areas where weight reduction is a priority, but they should not be relied upon as a primary radiation shield.

Modern Metal Alloys

Advancements in metallurgy have led to the development of modern metal alloys that offer improved strength, durability, and resistance to environmental factors. Alloys such as stainless steel, titanium, and specialized aluminium alloys are increasingly used in construction for their superior properties.

Advantages:

Strength and Durability: Modern metal alloys offer excellent strength and durability, making them suitable for high-stress applications.

Radiation Shielding: Metals, particularly those with high atomic numbers like lead or certain steel alloys, can provide good radiation shielding.

Corrosion Resistance: Many modern alloys are designed to resist corrosion, which is essential for long-term durability in harsh environments.

Limitations:

Cost: High-performance metal alloys can be expensive, particularly when used in large quantities.

Weight: The density of metals can make them heavy, which may require additional structural support in some applications.

Applications: Modern metal alloys can be used in the construction of critical structural components, such as doors, windows, and frames. They are also ideal for reinforcing walls or creating barriers in high-risk areas of a fallout shelter. In some cases, metals like stainless steel can be used for radiation shielding, particularly when combined with other materials.

Modern building materials offer a range of possibilities for enhancing the safety and effectiveness of fallout shelters. While some of these materials, like high-performance concrete and modern metal alloys, provide excellent radiation shielding, others, like SIPs and AAC, are better suited for insulation and structural support. By carefully evaluating the properties of these materials and understanding their strengths and limitations, you can make informed decisions about how to incorporate them into your shelter design. The key to effective fallout protection lies in combining these materials in a way that maximizes their protective qualities, ensuring that your shelter is well-equipped to withstand the challenges of a nuclear fallout scenario.

Medical Symptoms of Radiation Poisoning

Radiation poisoning, also known as Acute Radiation Syndrome (ARS), is a serious and potentially fatal condition caused by exposure to high doses of ionizing radiation over a short period. Understanding the medical symptoms of radiation poisoning is crucial for identifying the condition early, seeking appropriate medical care, and implementing measures to minimize further harm. This chapter delves into the symptoms of radiation poisoning, the stages of the condition, and the factors that influence its severity and progression.

Understanding Acute Radiation Syndrome (ARS)

Radiation poisoning occurs when the body is exposed to a large dose of ionizing radiation in a short time, typically as a result of a nuclear explosion, industrial accident, or medical treatment gone wrong. The severity of ARS depends on the total dose of radiation received, the type of radiation, and the duration of exposure. The dose of radiation is measured in Sieverts (Sv), with higher doses leading to more severe symptoms and a greater risk of death.

Stages of Radiation Poisoning

Radiation poisoning typically progresses through several stages, each with distinct symptoms. These stages are:

Prodromal Stage (Initial Stage)

Latent Stage

Manifest Illness Stage

Recovery or Death

Prodromal Stage (Initial Stage)

The prodromal stage occurs within hours to days after exposure to a high dose of radiation. The symptoms during this stage are often non-specific and can mimic other illnesses, making it challenging to diagnose ARS early without knowledge of radiation exposure.

Symptoms:

Nausea and Vomiting: These are often the earliest signs of radiation poisoning and can occur within minutes to hours after exposure. The severity and onset of nausea and vomiting can indicate the radiation dose received, with more severe and earlier onset symptoms suggesting higher exposure.

Diarrhoea: Diarrhoea, sometimes bloody, may follow nausea and vomiting, especially at higher radiation doses. This symptom is caused by damage to the lining of the gastrointestinal tract.

Headache: A common symptom in the prodromal stage, headaches can range from mild to severe, depending on the radiation dose.

Fatigue: A general feeling of weakness or exhaustion often accompanies the other symptoms during this stage.

Fever: A mild to moderate fever may develop as the body's immune response is compromised by the radiation.

The prodromal stage can last from a few hours to several days. The intensity and duration of symptoms during this stage are directly related to the radiation dose. Higher doses typically result in more severe symptoms and a shorter latency period before the next stage begins.

Latent Stage

Following the prodromal stage, there is often a period of apparent improvement known as the latent stage. During this time, symptoms may subside or disappear entirely, leading to a false sense of recovery. However, this stage is deceptive, as the body is undergoing significant internal damage, particularly to rapidly dividing cells in the bone marrow, gastrointestinal tract, and skin.

Duration: The latent stage can last anywhere from a few hours to several weeks, depending on the radiation dose. The length of this stage is inversely proportional to the dose received—higher doses result in shorter latent periods.

Manifest Illness Stage

The manifest illness stage is when the full effects of radiation poisoning become apparent. The symptoms during this stage are more severe and reflect the extensive damage done to the body's systems. The specific symptoms experienced during this stage depend on the dose of radiation and the organs most affected.

Symptoms:

Hematopoietic Syndrome: This occurs when radiation damages the bone marrow, leading to a severe reduction in blood cell production. Symptoms include anemia, bleeding (due to a lack of platelets), and infections (due to a low white blood cell count). This syndrome typically occurs at doses between 0.7 and 10 Sv.

Gastrointestinal Syndrome: This syndrome is caused by damage to the cells lining the gastrointestinal tract, leading to severe nausea, vomiting, diarrhea, and dehydration. It occurs at doses between 6 and 30 Sv and can be fatal due to the breakdown of the intestinal lining, which leads to severe infections and fluid loss.

Cardiovascular and Central Nervous System (CNS) Syndrome: At extremely high radiation doses (above 30 Sv), radiation poisoning can cause rapid and severe damage to the heart, blood vessels, and brain. Symptoms include severe headache, dizziness, confusion, loss of consciousness, and seizures. Death can occur within hours to days due to brain swelling, cardiovascular collapse, or multi-organ failure.

Skin Damage: Radiation can cause burns, blisters, and ulcers on the skin, particularly in areas directly exposed to radiation. These injuries can range from mild redness (erythema) to severe burns that require medical treatment.

The manifest illness stage is critical because it determines the overall prognosis for the individual. Those with less severe symptoms and a slower progression through this stage may recover with appropriate medical care, while those with more severe symptoms may face a higher risk of death.

Recovery or Death

The final outcome of radiation poisoning depends on the radiation dose, the effectiveness of medical treatment, and the individual's overall health. Recovery is possible for individuals who receive lower doses of radiation and appropriate medical care. However, recovery can take weeks to months, and long-term health effects, such as an increased risk of cancer, are common.

Recovery: Recovery from radiation poisoning involves the gradual healing of damaged tissues, restoration of blood cell counts, and management of infections and other complications. Survivors may require long-term medical monitoring and treatment for chronic health conditions resulting from radiation exposure.

Death: At higher radiation doses, or in cases where medical care is inadequate, death can occur due to severe damage to vital organs, overwhelming infections, or multi-organ failure. Death may occur within days to weeks of exposure, depending on the severity of the poisoning.

Factors Influencing the Severity of Radiation Poisoning

Several factors influence the severity of radiation poisoning and the likelihood of recovery:

Total Dose of Radiation: The higher the dose, the more severe the symptoms and the greater the risk of death.

Rate of Exposure: A high dose received in a short time is more dangerous than the same dose spread out over a longer period.

Type of Radiation: Different types of radiation (alpha, beta, gamma) have varying effects on the body, with gamma rays being the most penetrating and harmful.

Individual Health: Pre-existing health conditions, age, and overall physical condition can affect an individual's ability to survive radiation poisoning.

Availability of Medical Care: Prompt and effective medical treatment can significantly improve the chances of survival, particularly in managing infections, fluid loss, and organ damage.

Long-Term Health Effects

Survivors of radiation poisoning often face long-term health challenges. The most significant of these is an increased risk of developing cancer, particularly leukemia, thyroid cancer, and breast cancer. Radiation can also cause damage to the cardiovascular system, leading to an increased risk of heart disease and stroke. Additionally, radiation exposure can cause genetic damage, which may affect future offspring.

Radiation poisoning is a complex and potentially life-threatening condition that requires prompt recognition and treatment. Understanding the symptoms and stages of radiation poisoning is crucial for identifying the condition early and seeking appropriate medical care. While the prognosis depends on several factors, including the radiation dose and the availability of medical treatment, early intervention can improve the chances of survival and recovery. By being aware of the signs and symptoms of radiation poisoning, you can better prepare for and respond to a nuclear fallout event, ensuring the best possible outcome for yourself and your loved ones.

First Aid and Immediate Treatments for Radiation Exposure

In the aftermath of a nuclear event, immediate response to radiation exposure can be critical in reducing the severity of Acute Radiation Syndrome (ARS) and improving the chances of survival. While professional medical treatment is essential for those exposed to high doses of radiation, knowing how to administer first aid and immediate treatments can make a significant difference in the critical hours following exposure. This chapter outlines the steps you should take to provide first aid and initial treatment for radiation exposure, focusing on actions that can help mitigate the damage and support the body's recovery.

Assessing the Situation and Radiation Exposure

The first step in responding to radiation exposure is to assess the situation and the level of exposure. This involves determining whether the individual has been exposed to external radiation, contaminated with radioactive particles, or both. Understanding the type and severity of exposure will guide your response and help prioritize the necessary actions.

External Radiation Exposure: This occurs when a person is exposed to radiation from an external source, such as a nuclear explosion or radioactive materials. The severity of external radiation exposure depends on the dose received, which is influenced by the distance from the source, the duration of exposure, and the presence of any shielding.

Contamination with Radioactive Particles: This involves the deposition of radioactive particles on the skin, clothing, or within the body through inhalation or ingestion. Contamination can lead to both external and internal radiation exposure, with the potential to cause severe health effects if not promptly addressed.

Immediate Actions to Reduce Radiation Exposure

If you or someone else has been exposed to radiation, taking immediate action is crucial to reduce the dose and limit the damage. The following steps should be taken as soon as possible:

Get to Safety: Move away from the source of radiation or fallout to reduce further exposure. If possible, seek shelter in a building with thick walls, a basement, or another area that offers protection from radiation. The further you are from the source, and the more shielding you have, the lower your radiation dose will be.

Remove Contaminated Clothing: If radioactive particles have contaminated clothing, remove the clothing immediately to reduce the risk of further exposure. Carefully take off the clothing without shaking it to avoid spreading radioactive dust. Place the contaminated clothing in a sealed plastic bag and store it away from people and pets.

Decontaminate the Skin: Once the contaminated clothing has been removed, wash exposed skin thoroughly with soap and lukewarm water. Avoid scrubbing the skin harshly, as this can cause abrasions that may allow radioactive particles to enter the body. If water is scarce, use a moist cloth to wipe the skin, focusing on areas that were exposed to fallout.

Protect the Respiratory System: If you are in an area with airborne radioactive particles, cover your mouth and nose with a cloth, mask, or other improvised barrier to reduce inhalation of radioactive dust. If possible, move to an area with cleaner air or use an air filtration system.

Seek Shelter: After initial decontamination, sheltering indoors is critical to avoid further exposure. Stay inside, ideally in a basement or interior room, and keep windows and doors sealed. Ventilate the space only through filtered air if necessary, to prevent the entry of radioactive particles.

First Aid for Radiation Exposure Symptoms

After taking steps to reduce further exposure, it's important to address any symptoms of radiation poisoning that may be present. The severity of symptoms will depend on the dose of radiation received and the time since exposure. Here's how to provide first aid for common symptoms of radiation poisoning:

Nausea and Vomiting: These are common symptoms in the early stages of radiation poisoning. To manage nausea, keep the individual hydrated by providing small sips of water or an electrolyte solution. If vomiting occurs, encourage the person to drink fluids to prevent dehydration. Rest and keep the person in a comfortable, quiet environment.

Diarrhoea: Diarrhoea can lead to dehydration, so it's essential to replace lost fluids with water, oral rehydration solutions, or clear broths. Avoid giving the individual solid foods until the diarrhoea subsides. If diarrhoea is severe or persistent, seek medical attention as soon as possible.

Fever: A fever may indicate an immune response or an infection due to the body's reduced ability to fight pathogens after radiation exposure. Administer over-the-counter fever reducers, such as acetaminophen or ibuprofen, if available. Keep the person cool with damp cloths or a fan, and ensure they stay hydrated.

Skin Burns and Blisters: Radiation can cause burns and blisters, especially in areas directly exposed to fallout. Gently clean the affected area with mild soap and water, and cover it with a clean, dry dressing. Avoid breaking blisters, as this increases the risk of infection. If the burns are severe, seek medical care.

Fatigue and Weakness: Encourage rest and keep the individual comfortable. Provide fluids and light nourishment as tolerated. Fatigue is a common symptom of radiation poisoning and may persist for days or weeks.

Infections: Radiation exposure weakens the immune system, making the body more susceptible to infections. If the person shows signs of infection (e.g., redness, swelling, pus, or fever), clean the affected area and cover it with a sterile dressing. If antibiotics are available, administer them as directed, but seek medical advice if possible.

Internal Decontamination

If radioactive materials have been inhaled, ingested, or entered the body through wounds, internal decontamination may be necessary. This process typically requires medical intervention, but there are some immediate steps you can take:

Prevent Further Absorption: If ingestion of radioactive materials is suspected, discourage vomiting, as this can cause further harm. Activated charcoal may be used, if available, to bind radioactive particles and reduce absorption in the gastrointestinal tract.

Seek Medical Treatment: Internal contamination is serious and requires professional medical care. Chelation therapy, potassium iodide (for radioactive iodine), and other treatments may be necessary to remove or block the absorption of radioactive materials in the body. Seek medical help as soon as possible.

Potassium Iodide (KI) for Thyroid Protection

In the event of exposure to radioactive iodine, which is a common byproduct of nuclear fallout, taking potassium iodide (KI) can help protect the thyroid gland from absorbing the radioactive iodine. KI saturates the thyroid with stable iodine, reducing the uptake of radioactive iodine and lowering the risk of thyroid cancer.

Dosage:

The appropriate dosage of KI depends on age and weight. Always follow the recommended dosage guidelines provided by public health authorities or the product manufacturer.

KI should be taken as soon as possible after exposure to radioactive iodine. The effectiveness of KI decreases the longer you wait after exposure.

Limitations:

KI only protects the thyroid gland and does not protect against other forms of radiation exposure.

It should be taken only in the event of confirmed exposure to radioactive iodine and under the guidance of health authorities.

Monitoring and Long-Term Care

After providing first aid and immediate treatments, it's important to continue monitoring the individual for any worsening symptoms or new developments. Radiation poisoning can have delayed effects, so ongoing observation and care are crucial.

Monitor Vital Signs: Keep track of the person's temperature, pulse, and breathing rate. Any significant changes should be noted and reported to a healthcare provider if available.

Watch for Worsening Symptoms: Symptoms like persistent vomiting, severe diarrhea, uncontrolled bleeding, or confusion may indicate a serious progression of radiation poisoning and require urgent medical attention.

Plan for Medical Evacuation: If you are in an area with limited medical resources, plan for evacuation to a facility that can provide specialized care for radiation exposure. Ensure that the individual is stable and protected from further exposure during the evacuation process.

Prepare for Long-Term Health Monitoring: Survivors of radiation exposure will need long-term medical monitoring to detect and manage potential delayed effects, such as cancer or chronic health conditions. Establish a plan for regular medical check-ups and screenings.

In the event of radiation exposure, swift and effective first aid can make a significant difference in reducing the severity of symptoms and improving the chances of survival. By understanding the immediate steps to take, including decontamination, symptom management, and internal protection, you can provide critical care in the crucial hours following exposure. While professional medical treatment is essential for long-term recovery, knowing how to administer first aid and immediate treatments can help stabilize the individual and mitigate the initial impact of radiation poisoning.

Long-Term Medical Care for Radiation Sickness

Long-term medical care for radiation sickness is a crucial aspect of managing the aftermath of radiation exposure. While immediate treatments focus on mitigating the acute effects of radiation, long-term care addresses the ongoing health challenges that survivors may face, including chronic illnesses, increased cancer risk, and psychological impacts. This chapter explores the key components of long-term medical care for radiation sickness, highlighting the importance of monitoring, treatment, and support for those affected.

Understanding the Long-Term Effects of Radiation Exposure

Radiation exposure, particularly at high doses, can have lasting effects on the body. These effects may not be immediately apparent and can manifest months or even years after the initial exposure. The severity and nature of these long-term effects depend on several factors, including the total dose of radiation received, the type of radiation, the areas of the body affected, and the individual's overall health and age.

Common Long-Term Effects of Radiation Exposure:

Increased Risk of Cancer: One of the most significant long-term risks of radiation exposure is the increased likelihood of developing cancer. The types of cancer most commonly associated with radiation exposure include leukemia, thyroid cancer, breast cancer, lung cancer, and skin cancer.

Cardiovascular Disease: Radiation exposure can damage the heart and blood vessels, leading to an increased risk of heart disease, stroke, and other cardiovascular conditions.

Chronic Fatigue and Weakness: Many survivors of radiation exposure experience long-term fatigue and a general sense of weakness. This can be due to damage to the bone marrow, immune system, or other organs.

Cognitive Impairments: Exposure to high levels of radiation can cause neurological damage, leading to memory loss, difficulty concentrating, and other cognitive issues.

Fertility Issues: Radiation can affect the reproductive organs, leading to reduced fertility, menstrual irregularities, or even sterility in severe cases.

Genetic Mutations: Radiation exposure can cause genetic mutations that may not affect the individual directly but can be passed on to future generations, leading to birth defects or other genetic disorders.

Monitoring and Early Detection

Long-term care for radiation sickness begins with regular monitoring and early detection of potential health issues. Survivors of radiation exposure should undergo regular medical check-ups, including screenings for cancer, cardiovascular disease, and other conditions associated with radiation exposure. Early detection of these conditions is crucial for effective treatment and management.

Key Monitoring Practices:

Cancer Screenings: Regular screenings for cancers commonly associated with radiation exposure, such as leukemia and thyroid cancer, are essential. These screenings may include blood tests, imaging studies (e.g., CT scans or MRIs), and physical examinations.

Blood Tests: Routine blood tests can help monitor the health of the bone marrow and detect signs of anaemia, infection, or other blood-related conditions.

Cardiovascular Assessments: Regular check-ups should include assessments of heart health, including blood pressure monitoring, cholesterol levels, and, if necessary, imaging studies to detect cardiovascular damage.

Neurological Evaluations: Periodic neurological evaluations can help identify cognitive impairments or other neurological issues that may develop over time.

Fertility Assessments: For individuals of reproductive age, fertility assessments may be necessary to evaluate the impact of radiation on reproductive health.

Treatment Options for Long-Term Health Issues

If long-term health issues are identified, appropriate treatment and management strategies must be implemented. The specific treatments will depend on the condition diagnosed, the severity of the symptoms, and the overall health of the individual.

Cancer Treatment:

Chemotherapy and Radiation Therapy: Ironically, while radiation exposure can cause cancer, radiation therapy is also a common treatment for cancer. The goal is to target cancerous cells with controlled doses of radiation while minimizing damage to healthy tissue.

Surgery: In some cases, surgical removal of cancerous tumors may be necessary, particularly if the cancer is detected early and localized.

Immunotherapy: This treatment uses the body's immune system to fight cancer. Immunotherapy can be particularly effective for certain types of cancer, such as melanoma and some lung cancers.

Cardiovascular Disease Management:

Medications: Medications such as statins (to lower cholesterol), beta-blockers (to reduce blood pressure), and aspirin (to prevent blood clots) may be prescribed to manage cardiovascular disease.

Lifestyle Changes: Adopting a heart-healthy lifestyle, including a balanced diet, regular exercise, and smoking cessation, is critical for managing cardiovascular risk.

Surgical Interventions: In severe cases, surgical procedures such as angioplasty, stent placement, or bypass surgery may be necessary to restore blood flow to the heart.

Chronic Fatigue and Cognitive Impairments:

Rehabilitation Programs: Physical therapy and occupational therapy can help individuals regain strength, improve mobility, and manage chronic fatigue.

Cognitive Therapy: Cognitive-behavioral therapy (CBT) and other forms of cognitive rehabilitation can assist in managing memory loss, concentration difficulties, and other cognitive impairments.

Fertility Issues:

Fertility Treatments: Depending on the severity of the impact on reproductive health, fertility treatments such as in vitro fertilization (IVF) may be an option for those wishing to conceive.

Hormone Therapy: Hormone replacement therapy may be necessary for individuals experiencing hormonal imbalances or menopause-like symptoms due to radiation exposure.

Psychological Support and Counseling

The psychological impact of surviving radiation exposure should not be underestimated. Many individuals experience anxiety, depression, post-traumatic stress disorder (PTSD), and other mental health challenges as a result of their experience. Providing psychological support and counseling is a critical component of long-term care.

Support Services:

Individual Therapy: One-on-one counseling with a mental health professional can help individuals process their experiences, manage anxiety or depression, and develop coping strategies.

Support Groups: Joining support groups with other radiation survivors can provide a sense of community and shared understanding, which can be invaluable for emotional healing.

Family Counseling: Radiation exposure can impact the entire family, particularly if genetic mutations or fertility issues are involved. Family counseling can help address these challenges and strengthen family bonds.

Lifestyle Modifications for Long-Term Health

In addition to medical treatments and psychological support, adopting a healthy lifestyle is essential for managing the long-term effects of radiation exposure. Survivors should be encouraged to make positive lifestyle changes that can help mitigate the risk of developing chronic conditions and support overall well-being.

Key Lifestyle Modifications:

Healthy Diet: A diet rich in fruits, vegetables, whole grains, and lean proteins can support immune function, reduce inflammation, and lower the risk of cancer and cardiovascular disease. Limiting processed foods, red meat, and sugar is also recommended.

Regular Exercise: Engaging in regular physical activity, such as walking, swimming, or yoga, can improve cardiovascular health, boost mood, and combat chronic fatigue.

Smoking Cessation: Smoking increases the risk of cancer and cardiovascular disease, particularly for individuals already at higher risk due to radiation exposure. Quitting smoking is one of the most impactful lifestyle changes a survivor can make.

Stress Management: Chronic stress can weaken the immune system and contribute to the development of various health conditions. Techniques such as mindfulness meditation, deep breathing exercises, and relaxation therapy can help manage stress and promote mental health.

Regular Follow-Up and Health Monitoring

Long-term care for radiation sickness requires ongoing medical supervision and regular follow-up appointments to monitor for late-onset symptoms and manage chronic conditions. Survivors should work closely with their healthcare providers to establish a personalized care plan that includes:

Scheduled Screenings: Regular cancer screenings, cardiovascular assessments, and other health evaluations should be conducted based on the individual's risk factors and health status.

Vaccinations: Immunizations, such as the flu vaccine and pneumococcal vaccine, may be recommended to protect against infections in individuals with weakened immune systems.

Bone Density Monitoring: Radiation exposure can weaken bones, so monitoring bone density and taking steps to prevent osteoporosis may be necessary.

Social and Economic Support

Surviving radiation exposure can have long-term social and economic impacts, including job loss, financial strain, and social isolation. Access to social and economic support services is vital for helping survivors rebuild their lives and maintain their independence.

Support Services:

Disability Benefits: Survivors who are unable to work due to the long-term effects of radiation exposure may be eligible for disability benefits or other financial assistance programs.

Social Services: Access to housing assistance, food programs, and other social services can help alleviate the economic burden of long-term medical care.

Rehabilitation Programs: Vocational rehabilitation programs can assist survivors in retraining for new careers or finding employment that accommodates their health needs.

Long-term medical care for radiation sickness is a complex and multifaceted process that requires ongoing monitoring, treatment, and support. Survivors of radiation exposure face a range of health challenges, from increased cancer risk to cognitive impairments, and addressing these issues requires a comprehensive approach that includes medical treatment, psychological support, lifestyle modifications, and social services. By providing holistic care and support, healthcare providers, caregivers, and communities can help survivors manage the long-term effects of radiation exposure and lead fulfilling lives despite the challenges they face.

The Science of Fallout: Airborne and Settled Contamination

The science of fallout, particularly how it spreads and contaminates the environment, is crucial for understanding the risks associated with a nuclear event. Fallout consists of radioactive particles that are released into the atmosphere following a nuclear explosion. These particles can travel vast distances, settling on the ground, buildings, water sources, and even on people, leading to widespread contamination. This chapter delves into the science behind airborne and settled contamination, exploring how fallout spreads, the factors influencing its distribution, and the long-term implications of contamination.

What is Nuclear Fallout?

Nuclear fallout refers to the radioactive particles that are propelled into the upper atmosphere following a nuclear explosion. These particles are composed of fission products, which are the by-products of the nuclear reaction, as well as vaporized materials from the weapon and the environment around the explosion site. As these particles cool and condense, they begin to fall back to Earth, where they can contaminate everything they land on. Fallout particles vary in size, from microscopic dust to larger fragments, and they carry different types of radiation, including alpha, beta, and gamma radiation. The most dangerous of these is gamma radiation, which can penetrate deep into the body and cause severe damage to tissues and organs.

The Process of Fallout Formation

The formation of fallout begins at the moment of a nuclear explosion. When the bomb detonates, it releases an immense amount of energy, vaporizing the weapon and anything in its immediate vicinity. This material is carried upward by the heat of the explosion, forming a characteristic mushroom cloud that can reach the stratosphere, depending on the size of the explosion and the altitude at which it occurs. As the cloud rises and cools, the vaporized materials begin to condense into solid particles.

These particles are radioactive because they contain the fission products generated during the explosion. The size of these particles, and the height at which they form, determine how they will behave in the atmosphere. Larger, heavier particles tend to fall back to Earth relatively quickly, within hours of the explosion, and usually within a few hundred kilometers of the blast site. These particles are more likely to cause immediate, localized contamination. Smaller, lighter particles, however, can remain suspended in the atmosphere for longer periods and be carried over much greater distances by prevailing winds, leading to widespread contamination across entire regions or even continents.

Airborne Contamination

Airborne fallout is the primary concern immediately following a nuclear explosion. The radioactive particles that remain suspended in the atmosphere can be inhaled by people and animals, leading to internal exposure to radiation. Airborne particles can also settle on surfaces, where they pose a risk of external exposure to radiation.

Spread of Airborne Fallout:

Wind Patterns: The distribution of airborne fallout is heavily influenced by wind patterns. Winds at different altitudes can carry radioactive particles in different directions, dispersing them over wide areas. The jet stream, a fast-moving air current in the upper atmosphere, can transport fallout particles across entire continents within a matter of days.

Weather Conditions: Weather conditions, such as rain or snow, can cause fallout particles to be washed out of the atmosphere more quickly, leading to localized areas of intense contamination known as "hot spots." These hot spots can be particularly dangerous because they concentrate large amounts of radioactive material in a small area.

Altitude of Explosion: The altitude at which a nuclear explosion occurs also affects the spread of airborne fallout. A ground-level explosion will typically produce more fallout because it pulls in debris from the ground, which becomes irradiated and then dispersed. An airburst, on the other hand, tends to produce less fallout overall, but the particles it generates can be carried further by winds.

Health Risks of Airborne Contamination:

Inhalation: Inhaling radioactive particles is one of the most dangerous forms of exposure because it allows radiation to directly affect the lungs and other internal organs. This can lead to acute radiation poisoning and increase the risk of lung cancer and other respiratory conditions.

Skin Contact: Radioactive particles that settle on the skin can cause radiation burns, and if not washed off, they can lead to further contamination if they enter the body through wounds or ingestion.

Settled Contamination

As fallout particles settle back to Earth, they contaminate the ground, buildings, water sources, and vegetation. This settled contamination can persist for days, weeks, or even years, depending on the half-life of the radioactive isotopes involved. The half-life is the time it takes for half of the radioactive atoms in a sample to decay into a more stable form.

Factors Influencing Settled Contamination:

Particle Size: Larger particles tend to settle more quickly and closer to the explosion site, leading to higher levels of contamination in these areas. Smaller particles can be carried further and take longer to settle, leading to a more diffuse pattern of contamination.

Topography: The landscape can influence where fallout settles. Valleys, low-lying areas, and bodies of water can act as natural collection points for fallout, leading to higher concentrations of radioactive material in these areas.

Surface Type: Fallout particles can adhere to different surfaces in varying ways. Rough, porous surfaces, such as soil and vegetation, tend to retain fallout more effectively than smooth, hard surfaces like asphalt or concrete. This can make decontamination efforts more challenging in rural or forested areas compared to urban environments.

Health Risks of Settled Contamination:

Ingestion: One of the primary risks associated with settled fallout is the contamination of food and water supplies. Radioactive particles can settle on crops, infiltrate water sources, and be absorbed by livestock, leading to internal exposure when these foods are consumed. This can result in long-term health effects, including an increased risk of cancers and other diseases.

External Exposure: People living in or passing through contaminated areas can be exposed to radiation from settled fallout, particularly if they come into direct contact with contaminated soil, water, or objects. This exposure can lead to skin burns, radiation sickness, and an increased risk of cancer.

Long-Term Implications of Fallout Contamination

The long-term effects of fallout contamination depend on several factors, including the level of contamination, the half-life of the radioactive isotopes involved, and the effectiveness of decontamination efforts. Some isotopes, like iodine-131, have relatively short half-lives (about 8 days) but pose significant health risks if ingested shortly after a nuclear event. Others, like cesium-137 and strontium-90, have longer half-lives (about 30 years), meaning they can remain hazardous for decades.

Environmental Impact:

Soil Contamination: Fallout can contaminate soil, making it unsafe for agriculture and leading to long-term food insecurity. Contaminated soil may require removal or treatment to reduce radiation levels, a process that can be expensive and time-consuming.

Water Contamination: Fallout can infiltrate water sources, contaminating rivers, lakes, and groundwater. This can make water unsafe to drink and affect aquatic life, disrupting ecosystems and the food chain.

Vegetation and Wildlife: Plants and animals in contaminated areas can absorb radioactive particles, leading to bioaccumulation—the process by which radioactive materials become more concentrated as they move up the food chain. This can have serious implications for both the environment and human health, particularly in areas reliant on local agriculture and wildlife for food.

Decontamination Efforts:

Removal of Contaminated Soil: In areas with high levels of contamination, the top layer of soil may need to be removed and safely disposed of to reduce radiation levels.

Washing and Sealing Surfaces: Buildings and other structures can be decontaminated by washing surfaces with water and detergents or by sealing them with special coatings to prevent the spread of radioactive particles.

Water Treatment: Contaminated water sources may require treatment with filters, chemicals, or other methods to remove radioactive particles before the water is safe for consumption.

The science of fallout is complex, with airborne and settled contamination posing serious and long-lasting risks to both human health and the environment. Understanding how fallout spreads and settles is essential for planning effective protective measures, whether it's seeking shelter during the initial fallout, avoiding contaminated areas, or implementing long-term decontamination strategies. By comprehending the dynamics of fallout contamination, individuals and communities can better prepare for and respond to the challenges posed by a nuclear event, ultimately reducing the risks and ensuring a safer, more resilient future.

The Duration of Radioactive Danger: Understanding Half-Life

Understanding the concept of half-life is crucial for comprehending the duration of radioactive danger following a nuclear event. The half-life of a radioactive isotope is the time it takes for half of the radioactive atoms in a sample to decay into a more stable form. This decay process reduces the radioactivity of the material over time, but the length of time required for the danger to diminish depends on the specific isotopes involved. This chapter explores the concept of half-life, how it affects the persistence of radioactive contamination, and what this means for long-term safety and planning.

What is Half-Life?

Half-life is a fundamental concept in nuclear physics that describes the rate at which a radioactive substance decays. Each radioactive isotope has a characteristic half-life, ranging from fractions of a second to millions of years. During each half-life, the number of radioactive atoms in the material decreases by half. However, this decay does not happen uniformly; instead, it is a probabilistic process, meaning that while half of the atoms will have decayed by the end of the first half-life, the specific atoms that decay are random.

Key Points about Half-Life:

Not Linear: The decrease in radioactivity follows a logarithmic scale, not a linear one. After the first half-life, half of the original radioactive material remains. After the second half-life, a quarter remains, and so on.

Persistence of Danger: While radioactivity decreases over time, it can take many half-lives for a material to become safe. Even after several half-lives, a significant amount of radioactive material may still remain, posing a continued risk.

Varies by Isotope: Different isotopes have different half-lives, and the type of radiation they emit also varies. This affects how long the material remains hazardous and what kind of protection is necessary.

Common Radioactive Isotopes and Their Half-Lives

Several radioactive isotopes are commonly associated with nuclear fallout. Each has a different half-life, affecting how long it poses a danger after a nuclear event.

Iodine-131:

Half-Life: Approximately 8 days.

Radiation Type: Beta and gamma radiation.

Health Risks: Iodine-131 is particularly dangerous because it is readily absorbed by the thyroid gland, increasing the risk of thyroid cancer, especially in children and young adults.

Duration of Danger: Due to its short half-life, iodine-131 is most dangerous in the first few weeks after a nuclear event. However, its rapid decay means that its danger diminishes relatively quickly.

Cesium-137:

Half-Life: Approximately 30 years.

Radiation Type: Beta and gamma radiation.

Health Risks: Cesium-137 can be absorbed by the body, where it behaves like potassium, accumulating in muscles and other tissues. It poses long-term risks of cancer and other health issues.

Duration of Danger: Cesium-137 remains hazardous for decades, making it a long-term concern for contaminated areas.

Strontium-90:

Half-Life: Approximately 29 years.

Radiation Type: Beta radiation.

Health Risks: Strontium-90 behaves like calcium in the body, accumulating in bones and teeth, where it can cause bone cancer and leukemia.

Duration of Danger: Like cesium-137, strontium-90 poses long-term risks, with contamination lasting for decades.

Plutonium-239:

Half-Life: Approximately 24,100 years.

Radiation Type: Alpha radiation.

Health Risks: Plutonium-239 is highly toxic if inhaled or ingested, as alpha particles can cause significant damage to internal tissues. It poses a serious long-term health risk.

Duration of Danger: Due to its extremely long half-life, plutonium-239 remains hazardous for thousands of years, making it a persistent environmental and health threat.

Carbon-14:

Half-Life: Approximately 5,730 years.

Radiation Type: Beta radiation.

Health Risks: Carbon-14 is naturally occurring and less hazardous than other isotopes mentioned here, but it is used in radiocarbon dating and has long-term environmental implications.

Duration of Danger: Carbon-14's long half-life means it remains in the environment for millennia, though it is less of a direct health threat compared to more radioactive isotopes.

How Half-Life Affects Radiation Danger

The half-life of a radioactive isotope determines how long it remains a threat after a nuclear event. Isotopes with short half-lives, like iodine-131, pose an intense but brief danger, making immediate protection and sheltering critical in the days and weeks following exposure. On the other hand, isotopes with long half-lives, such as cesium-137 and plutonium-239, represent long-term hazards that can persist in the environment for years to millennia.

Short-Term vs. Long-Term Risks:

Short-Term: Isotopes with short half-lives decay quickly, meaning their radiation levels drop rapidly. However, during this time, they can cause significant damage if individuals are exposed without adequate protection. Immediate decontamination, sheltering, and avoiding contaminated food and water are essential.

Long-Term: Isotopes with long half-lives decay slowly, meaning they remain a threat for extended periods. These materials can contaminate soil, water, and food supplies, making areas unsafe for habitation or agriculture for decades or longer. Long-term strategies include decontamination, monitoring, and possibly abandoning heavily contaminated areas.

Practical Implications for Fallout Shelters and Safety

Understanding the half-life of radioactive isotopes helps inform the design and use of fallout shelters, as well as long-term safety measures:

Sheltering Duration: The need for prolonged sheltering depends on the half-life of the isotopes involved. For example, after a nuclear event, it's often advised to remain in a fallout shelter for at least 48 hours to avoid the most intense radiation from isotopes like iodine-131. However, for isotopes with longer half-lives, more extended sheltering or avoiding contaminated areas entirely may be necessary.

Decontamination Efforts: Knowing the half-life of contaminants helps prioritize decontamination efforts. For instance, areas contaminated with cesium-137 or strontium-90 require long-term decontamination and monitoring because these isotopes remain hazardous for decades.

Safe Re-entry and Habitation: Deciding when it's safe to reenter or inhabit a contaminated area depends on understanding the remaining levels of radioactivity. This decision is based on how much of the radioactive material has decayed and how much remains. For long-lived isotopes like plutonium-239, re-entry might not be feasible for many years.

Food and Water Safety: Radioactive contamination of food and water supplies can have long-lasting effects. For example, crops grown in soil contaminated with cesium-137 or strontium-90 can remain hazardous for decades. Long-term monitoring of agricultural areas and water sources is necessary to ensure they are safe for consumption.

Long-Term Environmental Impact

The environmental impact of radioactive contamination is closely linked to the half-life of the isotopes involved. Ecosystems in contaminated areas can suffer long-term damage, affecting wildlife, plant growth, and water quality. Understanding half-lives allows scientists and policymakers to predict how long these effects will last and what remediation efforts are necessary.

Ecosystem Recovery:

Short-Lived Isotopes: Areas contaminated by isotopes with short half-lives may recover more quickly, allowing for the eventual return of human habitation and agriculture.

Long-Lived Isotopes: Ecosystems contaminated by long-lived isotopes may require extensive remediation, and some areas may become uninhabitable for generations.

The half-life of radioactive isotopes is a critical factor in determining the duration of radioactive danger following a nuclear event. By understanding half-life, individuals and communities can make informed decisions about sheltering, decontamination, and long-term safety. Whether dealing with short-term threats from isotopes like iodine-131 or long-term hazards from cesium-137 and plutonium-239, knowledge of half-life is essential for protecting health and ensuring the safe use of contaminated areas in the future.

Contamination of Water Sources: Prevention and Treatment

Water is a vital resource, and in the event of a nuclear fallout, the contamination of water sources poses a significant threat to human health and survival. Radioactive particles from fallout can contaminate rivers, lakes, reservoirs, and groundwater, making the water unsafe for consumption and use. Understanding how to prevent and treat water contamination is crucial for ensuring access to safe drinking water in the aftermath of a nuclear event. This chapter explores the risks of water contamination, methods to prevent it, and techniques for treating contaminated water to make it safe for consumption.

How Water Sources Become Contaminated

Following a nuclear explosion, radioactive particles are dispersed into the atmosphere. These particles, known as fallout, can settle on the ground and directly contaminate water sources or be carried into water systems by rain, snow, or runoff. The contamination can affect both surface water, such as rivers and lakes, and groundwater, which includes wells and aquifers.

Types of Contaminants:

Alpha and Beta Particles: These particles can contaminate water directly or by adhering to soil particles that are washed into water sources.

Gamma Radiation: Gamma rays do not contaminate water directly but can pass through it, posing a risk if the water is exposed to a strong radiation source.

Long-Lived Isotopes: Isotopes like cesium-137 and strontium-90, which have long half-lives, can persist in the environment and continue to contaminate water sources for decades.

Risks of Drinking Contaminated Water

Consuming water contaminated with radioactive materials can lead to internal exposure to radiation, which is particularly dangerous. When radioactive isotopes are ingested, they can accumulate in specific organs or tissues, increasing the risk of cancers and other health problems over time.

Health Risks:

Thyroid Cancer: Iodine-131, if ingested, can accumulate in the thyroid gland, significantly increasing the risk of thyroid cancer.

Bone Cancer and Leukemia: Strontium-90 mimics calcium and can accumulate in bones and teeth, leading to bone cancer and leukemia.

Other Cancers: Cesium-137 can distribute throughout the body, increasing the risk of various cancers due to its ability to penetrate tissues and organs.

Preventing Water Contamination

Preventing contamination of water sources is a critical first step in ensuring access to safe drinking water after a nuclear event. While it may be impossible to entirely prevent fallout from contaminating water, certain precautions can minimize the risk.

Protective Measures:

Covering Water Sources: If feasible, covering wells, storage tanks, and other water sources can prevent radioactive particles from settling directly into the water. Tarps, plastic sheeting, or other protective covers can be used for this purpose.

Diverting Runoff: Creating barriers or drainage systems to divert runoff away from water sources can help reduce the amount of contamination that enters the water. This is particularly important in areas prone to heavy rainfall, which can carry fallout into rivers, lakes, and reservoirs.

Sealing Wells: Ensure that wells are properly sealed to prevent surface contamination from seeping into groundwater supplies. Well caps and seals should be inspected and reinforced as necessary.

Identifying Contaminated Water

After a nuclear event, it's essential to assume that all water sources may be contaminated until proven otherwise. The first step in dealing with potentially contaminated water is identifying the presence of radioactive materials.

Testing Water for Contamination:

Geiger Counters: While not specifically designed for water, Geiger counters can detect radiation in water samples, indicating the presence of contamination.

Laboratory Testing: For more accurate results, water samples should be sent to a laboratory equipped to measure specific isotopes, such as cesium-137, iodine-131, and strontium-90. This type of testing provides detailed information about the type and level of contamination.

Field Test Kits: Some field test kits can provide quick, albeit less precise, results for detecting specific radioactive isotopes in water. These kits can be useful for initial assessments.

Treating Contaminated Water

If water sources are found to be contaminated, treatment is necessary to make the water safe for consumption. There are several methods available for treating radioactive contamination in water, though each method has its limitations and may require additional equipment and resources.

Treatment Methods:

Filtration:

Activated Carbon Filters: These filters can remove some radioactive particles, particularly iodine-131, by adsorbing them onto the carbon. However, activated carbon filters are less effective against other isotopes like cesium-137 and strontium-90.

Reverse Osmosis: This method is highly effective at removing a wide range of contaminants, including radioactive particles. Reverse osmosis systems use a semi-permeable membrane to separate contaminants from water, making it one of the best options for treating contaminated water. However, these systems can be expensive and require a reliable power source.

Distillation:

Water Distillers: Distillation involves boiling water and then condensing the steam back into liquid form, leaving most contaminants behind. This process can effectively remove radioactive particles from water, although some volatile contaminants, such as tritium, may not be entirely removed. Distillation systems can be homemade or purchased commercially, but they require significant energy to operate.

Ion Exchange:

Ion Exchange Resins: These resins can remove certain radioactive isotopes, such as cesium-137 and strontium-90, from water by exchanging them with harmless ions. Ion exchange systems are effective but may need to be combined with other treatment methods to ensure complete decontamination. The effectiveness of ion exchange depends on the type of resin used and the specific contaminants present.

Chemical Treatments:

Potassium Iodide (KI): While not a treatment for water itself, potassium iodide can be taken as a prophylactic measure to protect the thyroid from radioactive iodine in contaminated water. KI does not purify water but helps protect the body from one specific contaminant.

Precipitation: Adding chemicals such as aluminum sulfate (alum) or ferric chloride to water can cause radioactive particles to precipitate out, forming a sediment that can be removed. This method is often used in conjunction with filtration to improve water quality.

Boiling:

Boiling Water: Boiling can help reduce some types of microbial contamination but is generally ineffective against radioactive particles. In some cases, boiling may concentrate radioactive materials if the water is not properly treated afterward.

Long-Term Strategies for Safe Water

While immediate treatment methods are crucial for making water safe in the short term, long-term strategies are necessary for ensuring continued access to uncontaminated water.

Long-Term Solutions:

Alternative Water Sources: Identifying and securing alternative water sources, such as deep aquifers that are less likely to be contaminated by surface fallout, can provide a long-term solution. This may involve drilling new wells or sourcing water from uncontaminated regions.

Water Storage: Storing clean water before a nuclear event can provide a reliable supply during the immediate aftermath. Large, sealed containers that are kept in a secure, covered location can protect water from contamination.

Rainwater Harvesting: Rainwater harvesting systems can be adapted to filter and purify water collected from rainfall, provided the rain itself is not contaminated. Initial runoff should be discarded to reduce contamination from roof surfaces, and the remaining water should be treated before use.

Ensuring Water Safety during Evacuation

In some cases, evacuation from a contaminated area may be necessary. Ensuring access to safe water during evacuation is critical for survival.

Evacuation Tips:

Portable Water Filters: Carrying portable water filters, such as those with activated carbon or reverse osmosis capabilities, can provide access to safe water on the move.

Water Purification Tablets: While primarily designed for microbial contamination, water purification tablets can be a useful addition to an evacuation kit. They should be used in conjunction with other methods for radioactive contamination.

Packaged Water: Bringing bottled or packaged water during evacuation ensures a reliable supply, particularly in the immediate aftermath of a nuclear event when other water sources may be compromised.

The contamination of water sources is one of the most critical challenges following a nuclear fallout event. Preventing and treating contaminated water is essential for maintaining health and survival. Long-term strategies, such as securing alternative water sources and effective water storage, further enhance resilience against the ongoing threat of radioactive contamination in water supplies.

Protecting Your Food Supply from Fallout

In the aftermath of a nuclear event, one of the most pressing concerns is ensuring that your food supply remains safe from radioactive contamination. Fallout can settle on crops, infiltrate food storage areas, and contaminate livestock, posing significant health risks if consumed. Protecting your food supply from fallout is crucial for long-term survival and health. This chapter explores the strategies and techniques you can use to safeguard your food, from initial fallout protection to long-term storage solutions.

Understanding How Fallout Contaminates Food

Fallout consists of radioactive particles that can settle on surfaces, including crops, soil, and water. When these particles come into contact with food, they can contaminate it, making it unsafe to eat. The degree of contamination depends on several factors, including the type of radioactive isotopes, the level of fallout, and the method of food storage.

Key Contaminants:

Surface Contamination: Fallout particles can settle on the surface of fruits, vegetables, and other food items. This type of contamination can often be mitigated through thorough washing or peeling.

Soil Contamination: Radioactive particles can infiltrate the soil, leading to long-term contamination of crops that grow in that soil. This type of contamination is more challenging to address, as it can affect the entire plant, not just the surface.

Water Contamination: Contaminated water can introduce radioactive particles into food through irrigation, food preparation, or contamination of water supplies used by livestock.

Airborne Contamination: Fallout particles in the air can directly contaminate food if it is left exposed. This is particularly a concern during and immediately after a fallout event.

Immediate Actions to Protect Your Food Supply

In the event of nuclear fallout, immediate action is necessary to protect your food supply from contamination. The following steps should be taken as soon as possible to minimize the risk of contamination:

Move Food Indoors: Any food stored outside, such as garden produce, should be moved indoors to protect it from airborne fallout. This includes food in outdoor storage units, on patios, or in open containers.

Cover Food Supplies: If it's not possible to move food indoors, cover it with tarps, plastic sheeting, or other protective materials to prevent fallout particles from settling on it. Make sure the covering is secure and does not allow particles to seep through.

Seal Food Storage Areas: Ensure that all food storage areas, such as pantries, cupboards, and refrigerators, are tightly sealed. Use duct tape, plastic sheeting, or other materials to seal gaps around doors and windows to prevent fallout from entering these spaces.

Protect Livestock Feed: If you have livestock, cover their feed and water sources to prevent contamination. Move feed indoors or cover it with protective materials, and consider providing clean, uncontaminated water from stored supplies.

Long-Term Food Storage Solutions

For long-term survival, it's important to have a food supply that is safe from contamination and capable of sustaining you and your family over an extended period. Here are some strategies for ensuring the safety and longevity of your food supply:

Storing Non-Perishable Foods:

Canned Goods: Canned foods are among the safest options for long-term storage as they are sealed and protected from contamination. Ensure that canned goods are stored in a cool, dry place, and check them regularly for signs of damage or corrosion.

Dry Foods: Store dry goods like rice, beans, pasta, and grains in airtight containers. Use Mylar bags, vacuum-sealed bags, or food-grade buckets with tight-fitting lids to prevent contamination. Consider using oxygen absorbers to extend shelf life.

Freeze-Dried Foods: Freeze-dried foods are lightweight, have a long shelf life, and are easy to store. They are also less likely to be contaminated because they are typically sealed in airtight, moisture-proof packaging.

Building a Fallout-Protected Pantry:

Underground Storage: An underground pantry or root cellar can provide natural protection from fallout. Ensure the space is well-sealed, with proper ventilation, and lined with materials that prevent contamination from seeping in.

Sealed Cabinets and Shelving: Inside your home, use sealed cabinets or shelving units to store food. Consider using plastic or metal storage bins that can be sealed with tight-fitting lids. Label all containers with the contents and date of storage.

Rotating Your Food Supply:

First In, First Out (FIFO): To maintain a fresh supply of food, practice the FIFO method, where you use the oldest food first and replace it with newer stock. This helps ensure that your food remains fresh and reduces the risk of spoilage.

Regular Inspections: Periodically inspect your food storage to check for signs of contamination, spoilage, or damage. Replace any compromised items and ensure that storage areas remain sealed and protected.

Gardening and Agriculture in a Fallout Scenario

If you rely on gardening or agriculture for food, it's important to take steps to protect your crops from fallout contamination. This involves both immediate actions during a fallout event and long-term strategies for maintaining a safe food supply.

Immediate Protection for Gardens:

Covering Crops: During a fallout event, cover crops with tarps, plastic sheeting, or row covers to prevent contamination. If contamination is expected, avoid harvesting crops until they can be properly tested or decontaminated.

Emergency Harvesting: If possible, harvest any mature crops before fallout arrives. Store them indoors or in a protected area to prevent contamination.

Long-Term Soil Management:

Soil Testing: After a fallout event, test your soil for radioactive contamination. This can help you determine whether it is safe to plant new crops or if decontamination measures are necessary.

Soil Remediation: If your soil is contaminated, consider remediation techniques such as removing the top layer of soil, adding clean topsoil, or using phytoremediation (growing plants that can absorb and remove radioactive particles).

Raised Beds and Containers: Growing food in raised beds or containers filled with uncontaminated soil can provide a safe alternative if your ground soil is compromised.

Watering and Irrigation:

Using Clean Water: Ensure that the water used for irrigation is free from contamination. If your primary water source is compromised, consider using stored water or collecting rainwater that has been filtered and tested.

Avoiding Contaminated Water Sources: Do not use water from potentially contaminated lakes, rivers, or reservoirs for irrigation until it has been tested and treated.

Decontaminating Food Supplies

If you suspect that your food supply has been exposed to fallout, it's essential to decontaminate it before consumption. While not all types of contamination can be removed, certain methods can reduce the risk.

Washing and Peeling:

Fruits and Vegetables: Wash fruits and vegetables thoroughly under running water to remove surface contamination. Use a brush to scrub the surface and consider peeling fruits and vegetables with thick skins. Discard the outer leaves of leafy vegetables.

Soaking: Soaking produce in a solution of water and baking soda or a specialized produce wash may help remove additional contaminants. Rinse thoroughly after soaking.

Cooking:

Boiling and Steaming: Boiling or steaming food can help reduce some forms of contamination, particularly if the water is changed during cooking. However, this method is not effective against all types of radioactive contamination, especially if the radiation has penetrated the food.

Avoiding Contaminated Foods:

Do Not Consume Suspect Food: If you are unsure whether food is contaminated and you do not have the means to test or decontaminate it effectively, it is safer to avoid consuming it. In a fallout situation, the risks of ingesting radioactive materials are severe, and it's better to rely on known safe supplies.

Protecting Livestock and Animal Feed

For those who raise livestock, protecting your animals and their feed from fallout is essential to ensure a continued supply of safe meat, milk, and eggs.

Sheltering Livestock:

Move Animals Indoors: If possible, move livestock to indoor shelters that are protected from fallout. Ensure that barns or shelters are sealed and provide clean bedding and feed.

Ventilation: Maintain adequate ventilation in livestock shelters, but ensure that air intakes are filtered or protected to prevent fallout particles from entering.

Protecting Feed and Water:

Covering Feed: Store feed indoors or cover outdoor feed supplies with tarps or plastic sheeting to prevent contamination. Use sealed containers for smaller quantities of feed.

Providing Clean Water: Ensure that livestock have access to clean, uncontaminated water. Use stored water or protected water sources, and avoid using potentially contaminated surface water for drinking or washing.

Protecting your food supply from fallout is a critical component of survival in the aftermath of a nuclear event. By understanding how fallout contaminates food and taking immediate and long-term measures to safeguard your food and water, you can significantly reduce the risk of exposure to radioactive materials. Proper storage, decontamination, and careful management of agriculture and livestock will help ensure that you have a safe and reliable food supply, even in the most challenging circumstances. Preparing in advance and staying vigilant in the face of potential contamination can make all the difference in maintaining health and safety during a fallout scenario.

Safe Gardening Practices in a Post-Fallout World

In a post-fallout world, gardening can be both a source of nourishment and a means of regaining a sense of normalcy. However, the presence of radioactive contamination in the environment poses significant challenges to safe food production. To ensure that your gardening practices yield safe and healthy produce, it is essential to understand how to mitigate the risks of contamination and implement strategies that protect your crops. This chapter explores safe gardening practices that can help you grow food in a contaminated environment while minimizing the risks to your health.

Understanding the Risks of Gardening after Fallout

The primary concern when gardening in a post-fallout world is the potential for radioactive contamination to affect your crops. Radioactive particles from fallout can settle on the soil, water, and plants, leading to both external and internal contamination of the produce. Consuming contaminated food can result in internal exposure to radiation, which poses serious health risks, including an increased likelihood of developing cancer and other long-term health problems.

Key Contaminants:

Surface Contamination: Fallout particles that settle on the surface of plants can be washed off, but the process must be thorough to reduce the risk effectively.

Soil Contamination: Radioactive particles can be absorbed by plants through their roots, leading to internal contamination that is more difficult to remove.

Water Contamination: Using contaminated water for irrigation can introduce radioactive particles to the soil and plants.

Initial Steps for Safe Gardening

Before starting or continuing your garden in a post-fallout environment, there are several crucial steps you should take to assess and mitigate the risks.

Soil Testing:

Test for Contamination: Conduct soil tests to determine the level of radioactive contamination present. This can be done using specialized testing kits or by sending soil samples to a laboratory. The results will help you decide whether the soil is safe for gardening or if remediation is necessary.

Identify Contaminants: Understand the specific radioactive isotopes present in your soil, as different isotopes have varying levels of danger and persistence. For example, cesium-137 and strontium-90 are long-lived and pose significant long-term risks.

Selecting a Safe Location:

Choose a Protected Area: If possible, select a gardening location that is sheltered from fallout, such as an area shielded by buildings or trees. An indoor or greenhouse garden may provide additional protection from airborne contaminants.

Consider Raised Beds: Raised beds filled with clean, uncontaminated soil can provide a safer alternative to planting directly in contaminated ground. These beds can be constructed from materials that block or reduce radiation exposure.

Water Testing and Safety:

Test Water Sources: Ensure that your water source is free from radioactive contamination before using it for irrigation. Contaminated water can introduce fallout particles into the soil and onto your plants.

Use Alternative Water Sources: If your primary water source is contaminated, consider using stored rainwater, filtered water, or water from a safe, uncontaminated well for your garden.

Soil Remediation Techniques

If your soil is contaminated, you will need to take steps to remediate it before planting. Soil remediation is the process of reducing or removing contaminants from the soil to make it safer for gardening.

Topsoil Removal:

Remove Contaminated Soil: In cases of severe contamination, removing the top few inches of soil can help reduce the level of radioactive particles. This contaminated soil should be carefully disposed of in a manner that prevents further environmental spread.

Add Clean Topsoil: After removing the contaminated layer, replace it with clean, uncontaminated topsoil. Ensure that this new soil is free from radioactive particles by sourcing it from a reliable, tested location.

Phytoremediation:

Plant Hyper-accumulators: Certain plants, known as hyper-accumulators, can absorb heavy metals and radioactive isotopes from the soil. Examples include sunflowers, mustard greens, and certain types of grasses. After these plants grow, they should be carefully removed and disposed of as hazardous waste, as they will contain concentrated levels of contaminants.

Repeated Cycles: Phytoremediation may need to be repeated over several growing cycles to significantly reduce contamination levels. This process can be time-consuming but may be necessary for long-term soil health.

Soil Amendments:

Use of Lime and Potassium: Adding lime and potassium to the soil can help reduce the uptake of certain radioactive isotopes, such as cesium-137, by plants. This can lower the overall level of contamination in the crops.

Organic Matter: Incorporating organic matter, such as compost, into the soil can improve its structure and health, potentially reducing the mobility of radioactive particles within the soil.

Planting and Growing in Contaminated Soil

If you must grow in soil that is still partially contaminated, there are several strategies you can use to reduce the risk of contamination in your crops.

Crop Selection:

Choose Low-Absorption Crops: Some plants absorb fewer radioactive particles than others. Leafy greens like spinach and lettuce are more likely to absorb contaminants, while fruits and tubers like tomatoes, potatoes, and squash tend to absorb less. Focus on planting crops that are less likely to accumulate radioactive isotopes.

Short-Season Crops: Crops with shorter growing seasons spend less time in the contaminated soil, reducing their exposure to radioactive particles. Consider planting vegetables that mature quickly, such as radishes, beans, and peas.

Growing Techniques:

Mulching: Apply a thick layer of mulch over the soil to prevent soil particles from splashing onto plants during watering or rain. Mulch also helps protect the soil from wind erosion, which can spread contamination.

Drip Irrigation: Use drip irrigation systems to minimize water contact with the plants' leaves and stems. This reduces the risk of contamination from both the soil and the water.

Barrier Protection: If possible, cover your garden with row covers, plastic sheeting, or netting to protect plants from airborne fallout and further contamination. These barriers can also help reduce the risk of contamination from rainfall.

Harvesting and Post-Harvest Handling:

Careful Harvesting: When harvesting crops, handle them carefully to avoid transferring contaminated soil onto the produce. Wear gloves and avoid direct contact with the soil.

Washing Produce: Thoroughly wash all harvested produce with clean water to remove any surface contamination. For leafy greens, consider multiple rinses and peeling away the outer layers to reduce the risk further.

Peeling and Cooking: Peel root vegetables and other crops with thick skins to remove any contamination that may have settled on the surface. Cooking can also help reduce the risk of ingesting certain contaminants, although it is not a guarantee of safety.

Long-Term Gardening Strategies

In a post-fallout world, gardening safely may require ongoing adjustments and strategies to ensure the health and safety of your crops and soil.

Regular Soil Testing:

Monitor Contamination Levels: Regularly test your soil for contamination to track changes over time and adjust your gardening practices accordingly. This ongoing monitoring is essential for ensuring that your soil remains safe for food production.

Crop Rotation and Diversity:

Rotate Crops: Practice crop rotation to reduce the buildup of contaminants in the soil and to maintain soil health. Avoid planting the same type of crop in the same location year after year.

Plant Diversity: Incorporate a diverse range of crops in your garden to spread the risk of contamination. Different plants absorb contaminants at different rates, so a diverse garden may help mitigate the overall risk.

Hydroponic Gardening:

Consider Hydroponics: Hydroponic gardening, where plants are grown in a nutrient-rich water solution rather than soil, can eliminate the risk of soil contamination altogether. This method requires an initial investment in equipment but can provide a safe and controlled environment for growing food.

Greenhouse Gardening:

Grow in a Greenhouse: Using a greenhouse can protect your garden from airborne contaminants and provide a controlled environment for growing food. Greenhouses can also extend your growing season and protect crops from harsh weather conditions.

Community and Support Networks

In a post-fallout world, collaborating with your community can enhance food security and safety. Sharing resources, knowledge, and labor can help ensure that everyone has access to safe, healthy food.

Community Gardens:

Establish Shared Gardens: Work with your community to establish shared gardens in safe locations. Community gardens can be more efficient and allow for collective soil remediation efforts, testing, and monitoring.

Seed Saving and Sharing:

Save and Share Seeds: Save seeds from your healthiest, least-contaminated crops for future planting. Sharing seeds with others can help maintain a diverse and resilient food supply.

Educational Programs:

Promote Education: Encourage local educational programs on safe gardening practices in a post-fallout environment. Knowledge sharing is crucial for collective survival and health.

Gardening in a post-fallout world presents significant challenges, but with careful planning, soil management, and protective measures, it is possible to grow safe, healthy food. By understanding the risks and implementing the strategies outlined in this chapter, you can create a garden that minimizes the dangers of radioactive contamination and supports long-term food security. Regular monitoring, soil remediation, and adaptive gardening practices will be essential as you navigate the complexities of growing food in a contaminated environment. Through resilience, knowledge, and community collaboration, you can maintain a sustainable and safe food supply even in the most challenging conditions.

How to Decontaminate Your Home after a Fallout Event

Decontaminating your home after a fallout event is a critical step in ensuring the safety and health of your family. Radioactive particles from fallout can settle on surfaces, infiltrate living spaces, and pose long-term risks if not properly addressed. Effective decontamination involves thorough cleaning, removal of contaminated materials, and ongoing monitoring to reduce radiation exposure. This chapter provides a step-by-step guide on how to decontaminate your home after a fallout event, focusing on practical techniques, safety measures, and long-term considerations.

Understanding the Nature of Fallout Contamination

Fallout consists of radioactive particles that can settle on any surface, including walls, floors, furniture, and personal items. These particles emit radiation, which can be harmful if ingested, inhaled, or if they come into prolonged contact with the skin. The goal of decontamination is to remove or neutralize these particles to reduce radiation exposure inside your home.

Types of Contamination:

Surface Contamination: Fallout particles that settle on surfaces like countertops, floors, and walls. These are often the easiest to remove.

Embedded Contamination: Particles that have penetrated porous materials like fabric, carpet, or wood. This contamination can be more challenging to address and may require more aggressive cleaning or removal.

Airborne Contamination: Fallout particles that remain suspended in the air and can settle over time. Ventilation and air filtration are key to reducing airborne contamination.

Immediate Steps after a Fallout Event

Before beginning the decontamination process, it's crucial to take immediate steps to prevent further contamination and protect yourself from exposure.

Seal Your Home:

Close All Openings: Immediately after a fallout event, close all windows, doors, vents, and other openings to prevent additional fallout particles from entering your home.

Seal Gaps: Use duct tape, plastic sheeting, or other materials to seal gaps around doors, windows, and vents. This will help minimize the entry of contaminated air.

Protect Yourself:

Wear Protective Gear: Before starting decontamination, wear protective clothing, including long sleeves, gloves, goggles, and a mask (preferably an N95 respirator) to reduce your exposure to radioactive particles.

Avoid Eating or Drinking: Do not eat, drink, or touch your face while decontaminating to prevent ingestion of radioactive particles.

Ventilate with Caution:

Filtered Ventilation: If possible, ventilate your home using filtered air to reduce airborne particles. Use air purifiers with HEPA filters to capture fallout particles from the air.

Decontaminating Interior Surfaces

Once you've taken initial precautions, you can begin decontaminating the interior surfaces of your home. The goal is to remove as much radioactive material as possible while minimizing your exposure.

Cleaning Hard Surfaces:

Wipe Down Surfaces: Use damp cloths or disposable wipes to clean hard surfaces such as countertops, tables, walls, and floors. Start from the highest surfaces and work your way down to avoid re-contaminating cleaned areas.

Use Detergent Solutions: A solution of water and household detergent can help lift radioactive particles from surfaces. Avoid using abrasive cleaners that might spread contamination.

Dispose of Cleaning Materials: After wiping down surfaces, place used cloths, wipes, and gloves in sealed plastic bags and dispose of them safely. Do not shake out cloths or rags, as this can spread contamination.

Vacuuming:

Use a HEPA Vacuum: If you need to vacuum, use a vacuum cleaner equipped with a HEPA filter to capture fine radioactive particles. Avoid using regular vacuums, as they can release particles back into the air.

Dispose of Vacuum Bags Carefully: After vacuuming, remove and seal the vacuum bag or filter in a plastic bag and dispose of it as contaminated waste.

Decontaminating Fabrics and Upholstery:

Launder Washable Fabrics: Wash clothing, bedding, curtains, and other washable fabrics in hot water with a strong detergent. Repeat the washing process multiple times to remove as much contamination as possible.

Steam Cleaning: For carpets and upholstered furniture, consider using a steam cleaner, which can help extract embedded particles. Follow up with a thorough vacuuming using a HEPA vacuum.

Dispose of Contaminated Items: If items are heavily contaminated and cannot be effectively cleaned, consider disposing of them. Place them in sealed plastic bags and label them as contaminated.

Cleaning Personal Items:

Wipe Down Electronics: Carefully wipe down electronic devices and other personal items with damp cloths or wipes. Be cautious not to damage sensitive equipment.

Disinfecting: Use a disinfectant spray or wipes to clean items that are frequently handled, such as remote controls, phones, and keys.

Decontaminating Exterior Surfaces

Exterior surfaces of your home, including walls, roofs, and walkways, can also become contaminated with fallout. Decontaminating these areas reduces the risk of bringing contamination into the home.

Rinse and Wash Exterior Surfaces:

Use Water and Detergent: If possible, use a hose to rinse down exterior walls, windows, and roofs. Adding a detergent solution can help lift and remove fallout particles.

Pressure Washing: For more thorough cleaning, consider using a pressure washer with detergent. Be mindful of where the runoff goes, as it could contaminate the soil or water supply.

Cleaning Outdoor Areas:

Clean Walkways and Driveways: Sweep or hose down walkways, driveways, and patios to remove fallout particles. Use a broom to push debris away from the home and avoid spreading contamination.

Rake and Bag Debris: If your yard is contaminated, rake up leaves, grass clippings, and other debris, and place them in sealed bags for disposal. Avoid composting contaminated organic matter.

Dealing with Roof Contamination:

Roof Cleaning: If fallout has settled on your roof, it's important to clean it to prevent contamination from washing into gutters and downspouts. Use a hose or pressure washer to rinse the roof, directing runoff away from the home and into a safe area.

Safe Disposal of Contaminated Materials

Proper disposal of contaminated materials is crucial to prevent further spread of radiation and protect public health.

Seal and Label Waste:

Use Heavy-Duty Bags: Place contaminated materials, such as cleaning rags, vacuum bags, and removed topsoil, in heavy-duty plastic bags. Double-bag if necessary to prevent leaks.

Label as Contaminated: Clearly label all bags and containers as containing radioactive materials. This will help ensure they are handled appropriately during disposal.

Follow Local Guidelines:

Disposal Instructions: Follow local guidelines and regulations for the disposal of radioactive waste. This may involve contacting local authorities or waste management services to arrange for safe disposal.

Ongoing Monitoring and Maintenance

Decontamination is not a one-time task; ongoing monitoring and maintenance are necessary to ensure that your home remains safe over time.

Regular Testing:

Use Radiation Detectors: Regularly use a Geiger counter or other radiation detection devices to monitor radiation levels in your home. Focus on high-risk areas, such as entrances, windows, and air vents.

Re-evaluate Contamination Levels: Periodically reassess the contamination levels in and around your home, especially after heavy rains or winds that could bring new fallout particles into your area.

Ventilation and Air Purification:

Air Filters: Continue using air purifiers with HEPA filters to reduce airborne contamination. Replace filters regularly according to the manufacturer's instructions.

Maintain Seals: Check the seals around doors, windows, and vents to ensure they remain intact and effective. Re-seal any gaps or cracks that develop over time.

Safe Living Practices:

Shoes Off Indoors: Adopt a shoes-off policy inside the home to reduce the tracking of contaminated particles from outside.

Routine Cleaning: Implement a routine cleaning schedule that includes wiping down surfaces, vacuuming with a HEPA vacuum, and washing fabrics to keep contamination at bay.

Decontamination of Water Supply

If your water supply has been contaminated by fallout, it's critical to take steps to decontaminate it or find an alternative water source.

Water Filtration and Purification:

Use Advanced Filters: Install water filters that can remove radioactive particles. Reverse osmosis systems and activated carbon filters are effective options.

Boiling Water: Boiling water can reduce biological contaminants but is not effective for removing radioactive particles. Use this method in combination with filtration.

Alternative Water Sources:

Stored Water: If possible, rely on stored water supplies until your primary source is safe. Ensure that stored water is kept in sealed, uncontaminated containers.

Rainwater Collection: If collecting rainwater, use a filtration system to remove contaminants before drinking or using the water for household purposes.

Psychological and Emotional Considerations

The process of decontaminating your home after a fallout event can be physically demanding and emotionally taxing. It's important to address the psychological impacts and take care of your mental health during this time.

Manage Stress:

Take Breaks: Decontaminating your home is a marathon, not a sprint. Take regular breaks to rest and avoid burnout.

Practice Mindfulness: Techniques like practicing mindfulness, deep breathing exercises, or meditation can help manage stress and anxiety during the decontamination process. These techniques can provide mental clarity and help you stay focused on the tasks at hand.

Seek Support:

Connect with Others: Reach out to family, friends, or neighbors who may also be dealing with fallout decontamination. Sharing experiences and advice can provide emotional support and practical solutions.

Professional Help: If the stress and anxiety become overwhelming, consider seeking professional mental health support. Talking to a counsellor or therapist can help you process the trauma and develop coping strategies.

Involve the Family:

Communicate Openly: If you have family members or housemates, communicate openly about the decontamination process, the reasons for safety precautions, and the importance of working together. This can help reduce fear and uncertainty.

Assign Tasks: Involve everyone in age-appropriate tasks during the decontamination process. Giving each person a role can create a sense of purpose and teamwork, helping to alleviate feelings of helplessness.

Focus on Progress:

Set Small Goals: Break the decontamination process into smaller, manageable tasks, and celebrate each completed step. This can help maintain motivation and provide a sense of accomplishment.

Keep a Log: Document your progress with a journal or checklist. Seeing the progress you've made can boost morale and provide reassurance that you're making your home safer.

Long-Term Considerations and Maintenance

Even after the initial decontamination is complete, maintaining a safe living environment in a post-fallout world requires ongoing vigilance and effort.

Regular Maintenance:

Inspect and Repair: Regularly inspect your home for any new signs of contamination or damage that could compromise its safety. This includes checking seals, air filters, and protective coverings.

Continue Decontamination Practices: Periodically repeat decontamination procedures, especially after natural events like heavy rain, windstorms, or snow that could introduce new fallout particles.

Preparedness for Future Events:

Stock Up on Supplies: Maintain a well-stocked supply of decontamination materials, including cleaning supplies, protective gear, and radiation detection equipment. Being prepared will help you respond quickly to any future contamination.

Plan for Evacuation: If your home becomes too contaminated to safely decontaminate, have an evacuation plan in place. Know where you would go, how you would get there, and what essentials to bring with you.

Community Involvement:

Engage with the Community: Work with your local community to develop or participate in collective decontamination efforts. Community-wide initiatives can address larger environmental issues, such as contaminated public spaces or water supplies.

Advocate for Safety Measures: Advocate for local government support in managing fallout contamination, including public health advisories, cleanup efforts, and access to decontamination resources.

Decontaminating your home after a fallout event is a critical process that requires thoroughness, caution, and patience. Remember that decontamination is not a one-time task; it involves ongoing monitoring, regular maintenance, and a commitment to safety. With careful planning, the right tools, and a supportive community, you can effectively manage the challenges of living in a post-fallout world and protect your home from the dangers of radioactive contamination.

Designing a Fallout Shelter: Key Considerations

Designing a fallout shelter is a crucial step in preparing for the possibility of a nuclear event. A well-designed shelter can provide protection from radioactive fallout, ensuring the safety and survival of you and your loved ones during and after a nuclear incident. This chapter outlines the key considerations for designing an effective fallout shelter, covering aspects such as location, structure, ventilation, supplies, and long-term survivability.

Location: Choosing the Right Spot

The location of your fallout shelter is one of the most important decisions in the design process. The ideal location will provide maximum protection from radiation and other hazards while being accessible and practical for your needs.

Key Considerations:

Underground vs. Aboveground:

Underground Shelters: Generally offer the best protection from radiation because the earth acts as a natural shield. Basements, converted root cellars, or purpose-built underground bunkers are common choices.

Aboveground Shelters: May be necessary if underground construction is not feasible due to high water tables or soil conditions. In this case, thick walls made of concrete, earth, or other dense materials are essential for radiation shielding.

Proximity to Home: Ideally, the shelter should be close enough to your home to reach quickly in an emergency but far enough to avoid direct damage from the blast. If incorporating a shelter within your home, a basement or specially reinforced room may be suitable.

Accessibility: Ensure that the shelter is easily accessible for all family members, including those with mobility challenges. Consider multiple access points if possible, to prevent entrapment in case one entrance becomes blocked.

Structural Integrity: Building a Safe Shelter

The structure of your fallout shelter must be robust enough to withstand not only radiation but also potential blast effects, such as shockwaves, debris, and fire. The materials used in construction play a crucial role in providing effective shielding.

Key Considerations:

Material Selection:

Concrete: One of the best materials for fallout shelters due to its density and availability. Thick concrete walls (at least 12-24 inches) provide excellent radiation protection.

Earth: Earth-bermed or earth-covered shelters are also effective, with soil thickness of at least 3 feet offering significant radiation shielding.

Steel and Lead: These materials can be used to reinforce walls or ceilings, though they are typically more expensive and require careful handling due to their weight and potential health risks.

Wall and Ceiling Thickness: The thicker the walls and ceiling, the better the protection. Aim for a minimum of 12 inches of concrete or its equivalent in other materials. The ceiling should be similarly reinforced, especially in aboveground shelters.

Blast Protection: In areas closer to potential nuclear targets, consider incorporating blast-resistant features, such as reinforced doors, blast valves for ventilation, and secure foundations. The shelter should be designed to withstand shockwaves and overpressure from a nearby explosion.

Ventilation and Air Filtration: Ensuring Safe Air Quality

Maintaining breathable air in a fallout shelter is critical, especially if you need to stay inside for an extended period. Proper ventilation and air filtration systems are essential to prevent the buildup of carbon dioxide, remove airborne contaminants, and ensure a supply of clean air.

Key Considerations:

Ventilation Systems:

Manual Ventilation: A hand-cranked ventilation system can ensure air circulation without relying on electricity. It should be easy to operate and maintain.

Powered Ventilation: If electricity is available, powered fans can help circulate air more effectively. However, ensure that there is a backup manual system in case of power failure.

Air Filtration:

HEPA Filters: High-Efficiency Particulate Air (HEPA) filters are essential for removing radioactive particles from the air. Ensure the filter system is properly installed and maintained.

Activated Carbon Filters: These can remove chemical contaminants and gases, providing additional protection.

Air Intake and Exhaust: Proper placement of air intake and exhaust vents is crucial. These should be shielded and equipped with blast valves to prevent contamination or damage from shockwaves.

Water and Sanitation: Managing Essential Needs

Water is a critical resource in a fallout shelter, both for drinking and for sanitation purposes. Ensuring a safe and reliable water supply, along with effective waste management, is key to long-term survivability.

Key Considerations:

Water Supply:

Stored Water: Plan to store enough water for all occupants for at least two weeks, with a minimum of 1 gallon per person per day. Water should be stored in clean, sealed containers and rotated regularly.

Water Filtration: Equip your shelter with a water filtration system capable of removing radioactive particles and other contaminants from any water sources you may need to use.

Rainwater Harvesting: If possible, design the shelter to collect and filter rainwater for additional supply, though this should be considered supplementary due to potential contamination risks.

Sanitation:

Toilets: A composting toilet or chemical toilet is ideal for a fallout shelter, as these do not require water and can handle waste safely. Ensure there is enough capacity or a plan for waste disposal.

Hygiene: Store sufficient hygiene supplies, such as soap, disinfectants, and personal care items. Wet wipes and hand sanitizer can be useful for conserving water.

Food Storage: Sustaining Life over Time

Your fallout shelter should be equipped with enough food to last at least two weeks, with provisions for longer stays if necessary. Food storage requires careful planning to ensure that it remains safe and nutritious over time.

Key Considerations:

Non-Perishable Foods: Focus on storing non-perishable foods with a long shelf life, such as canned goods, freeze-dried meals, rice, beans, and pasta. Consider nutritional balance and include a variety of foods to avoid diet fatigue.

Caloric Needs: Ensure that your food supply meets the caloric and nutritional needs of all shelter occupants. Plan for at least 2,000 calories per person per day.

Food Storage Conditions: Store food in a cool, dry place, protected from moisture and pests. Use airtight containers and rotate your food supply regularly to maintain freshness.

Cooking and Preparation: If possible, equip your shelter with a means of cooking, such as a camp stove or portable burner, along with appropriate fuel supplies. Be mindful of ventilation when using cooking equipment.

Power and Lighting: Maintaining Functionality

While it's possible to survive in a fallout shelter without power, having a reliable source of electricity can greatly improve comfort and functionality. Lighting, communication, and essential systems may require power.

Key Considerations:

Power Sources:

Generators: A generator can provide electricity for your shelter, but it requires a reliable fuel supply and proper ventilation to prevent carbon monoxide buildup.

Solar Power: Solar panels combined with battery storage can offer a renewable power source, though effectiveness depends on location and shelter design.

Battery Backup: Ensure you have a supply of batteries for flashlights, radios, and other essential devices. Rechargeable batteries and a hand-crank charger are also useful.

Lighting:

LED Lights: LED lights are energy-efficient and long-lasting, making them ideal for a fallout shelter. Consider both overhead lighting and portable lanterns or flashlights.

Emergency Lighting: Have backup emergency lighting options, such as glow sticks, candles, or hand-crank lanterns, in case your primary power source fails.

Communication and Monitoring: Staying Informed

Staying informed during a nuclear event is crucial for making decisions about when it's safe to leave the shelter and what actions to take. Communication devices and monitoring equipment can keep you connected and aware of your surroundings.

Key Considerations:

Radios:

Emergency Radio: A battery-powered or hand-crank emergency radio can receive weather alerts, news updates, and official instructions. Look for a model with AM/FM and NOAA weather channels.

Two-Way Radios: Two-way radios allow for communication with others, especially if cell service is disrupted. Ensure you have extra batteries or a means to recharge them.

Radiation Detection:

Geiger Counter: A Geiger counter or dosimeter is essential for monitoring radiation levels inside and outside the shelter. Regularly check radiation levels to determine when it is safe to leave the shelter.

Personal Dosimeters: Equip each shelter occupant with a personal dosimeter to track individual radiation exposure over time.

Psychological and Emotional Considerations: Maintaining Mental Health

Life in a fallout shelter can be stressful and challenging, especially during prolonged stays. Addressing psychological and emotional needs is just as important as meeting physical needs.

Key Considerations:

Comfort and Entertainment:

Books and Games: Stock the shelter with books, games, puzzles, and other entertainment options to help pass the time and reduce stress.

Comfort Items: Include personal comfort items, such as blankets, pillows, and familiar objects that can help maintain a sense of normalcy and emotional well-being.

Communication and Social Interaction:

Stay Connected: Regular communication with family members or other shelter occupants is crucial for maintaining morale. Plan for group activities or discussions to foster a sense of community.

Mental Health Resources: Consider including mental health resources, such as self-help books, journals, or recorded relaxation exercises. If possible, designate a quiet space for meditation or stress relief.

Long-Term Survival: Planning for Extended Stays

While fallout shelters are typically designed for short-term stays of a few days to a few weeks, it's important to consider the possibility of an extended stay if conditions outside remain hazardous. Planning for long-term survival involves ensuring that your shelter can support life for an extended period, potentially several months.

Long-Term Survival: Planning for Extended Stays

Key Considerations:

Extended Food and Water Supplies:

Food Production: If your shelter has enough space and the necessary resources, consider setting up a small indoor garden for growing vegetables, herbs, or sprouts. Hydroponic systems or container gardening can be viable options in a confined space.

Water Replenishment: Plan for methods to replenish your water supply, such as rainwater harvesting or water purification systems that can treat additional water sources. Storing extra water purification tablets or filters can also be beneficial.

Waste Management:

Expanded Sanitation Solutions: Over time, waste disposal becomes a critical issue. Composting toilets or waste incinerators can be used to manage human waste safely. Ensure that waste is handled in a way that prevents contamination of your living space and water supply.

Gray Water Management: Consider how to handle gray water (wastewater from washing and cleaning). If space and resources allow, set up a filtration or recycling system for gray water to conserve your clean water supply.

Health and Medical Care:

Medical Supplies: Stock an extensive first-aid kit with supplies for both immediate injuries and long-term health needs. Include a variety of medications, antiseptics, bandages, splints, and any prescription medications needed by shelter occupants.

Hygiene: Ensure that hygiene supplies are plentiful, including soap, disinfectants, toothbrushes, toothpaste, feminine hygiene products, and other essentials. Proper hygiene is crucial to prevent illness in a confined space.

Health Monitoring: Regularly monitor the health of all shelter occupants, and be prepared to address issues like respiratory problems, infections, and mental health challenges.

Physical Fitness and Exercise:

Exercise Space: Include space and equipment for physical exercise. Even in a small shelter, activities like stretching, yoga, or bodyweight exercises can help maintain physical health and mental well-being.

Routine: Establish a daily routine that includes time for physical activity. Staying active is important for both physical health and stress management.

Energy and Power Solutions:

Renewable Energy Sources: For long-term sustainability, consider renewable energy options like solar panels or wind turbines, paired with battery storage to ensure a continuous power supply.

Fuel Storage: If relying on generators, ensure you have a substantial and safely stored supply of fuel. Plan for how to manage fuel use efficiently to extend the power supply.

Resource Conservation:

Rationing: Implement a rationing system for food, water, and other essential supplies from the start. Carefully monitor consumption to avoid running out of critical resources.

Reuse and Recycling: Practice reuse and recycling of materials wherever possible. For example, containers, packaging, and even waste products may have secondary uses in a resource-limited environment.

Social and Emotional Well-being:

Community Building: If sheltering with others, focus on maintaining positive relationships and clear communication. Regular group discussions, shared meals, and cooperative activities can help build a sense of community and support.

Spiritual and Mental Health: Provide for spiritual needs by including religious texts, symbols, or other spiritual items as appropriate. Maintaining a sense of purpose and hope is crucial in a long-term survival situation.

Exit Strategy: Knowing When and How to Leave

Finally, designing a fallout shelter includes planning for an eventual exit. Knowing when and how to leave the shelter safely is critical, as leaving too early or unprepared can result in exposure to dangerous levels of radiation.

Key Considerations:

Monitoring Radiation Levels:

Regular Checks: Use radiation detection equipment to regularly check the radiation levels outside the shelter. Keep a log of readings to track trends and determine when it might be safe to exit.

Safe Levels: Familiarize yourself with the safe levels of radiation for re-entry. Typically, levels below 0.1 rem per hour (or 1 millisievert per hour) are considered safe for short-term exposure, but it is crucial to consult up-to-date guidelines.

Gradual Re-entry:

Initial Exits: When radiation levels have dropped to safe levels, make initial short excursions outside the shelter to assess conditions. Wear protective clothing and use a dosimeter to monitor exposure during these trips.

Shelter Reintegration: If the environment is safe enough for longer stays outside the shelter, begin reintegrating the shelter into your daily life by using it for rest or storage while gradually spending more time outside.

Post-Exit Safety:

Decontamination: After leaving the shelter, decontaminate any clothing, gear, or surfaces that were exposed to the outside environment before bringing them back into the shelter or home.

Long-Term Habitat: Once it's safe to leave the shelter permanently, focus on reestablishing a long-term, sustainable living situation. This may involve repairing and reinforcing your home, continuing decontamination efforts, and rebuilding a sense of normalcy.

Designing a fallout shelter involves careful consideration of multiple factors, from location and structure to ventilation, supplies, and long-term survivability. By addressing these key areas, you can create a shelter that not only protects you and your loved ones from the immediate dangers of nuclear fallout but also supports your well-being during an extended stay. A well-planned shelter is an investment in your safety, offering peace of mind in uncertain times and the best possible chance of survival in the event of a nuclear disaster.

Selecting the Right Location for Your Shelter

Selecting the right location for your fallout shelter is one of the most critical decisions in the design and construction process. The location will significantly impact the shelter's effectiveness in protecting you and your loved ones from radiation, blast effects, and other hazards associated with a nuclear event. This chapter will guide you through the key factors to consider when choosing the ideal location for your fallout shelter, whether you're building it underground, aboveground, or retrofitting an existing structure.

Understanding the Purpose of Your Shelter

Before selecting a location, it's essential to clearly define the purpose of your fallout shelter. Are you primarily seeking protection from radiation, or are you also concerned about blast effects, chemical or biological threats, or natural disasters? The purpose of the shelter will influence the type of location that is most suitable.

Key Considerations:

Radiation Protection: The primary goal of a fallout shelter is to shield you from radiation. This requires dense materials like earth or concrete to block harmful gamma rays and other radiation.

Blast Protection: If you are near a potential nuclear target, such as a military base or major city, blast protection becomes critical. In such cases, the location must also be shielded from shockwaves and debris.

Accessibility and Comfort: Consider how easily you can access the shelter in an emergency and how comfortable it will be for an extended stay. A location that is difficult to reach or that lacks basic amenities might not be ideal.

Proximity to Your Home

The shelter should be close enough to your primary living space to allow for rapid access in the event of a nuclear warning. However, proximity to your home also involves considering the safety of the shelter in relation to potential hazards nearby.

Key Considerations:

Immediate Access: The shelter should be located within a few minutes' walk from your home, ideally connected directly via a basement or an underground passage. This allows for quick entry in case of sudden fallout.

Avoiding Hazards: Ensure the shelter is not too close to hazards such as large windows, flammable materials, or structures that could collapse during a blast. Additionally, avoid areas prone to flooding, as water can compromise the shelter's integrity and safety.

Discreet Location: A shelter that is somewhat hidden or not immediately obvious can help protect against potential looting or unwanted attention in a post-disaster scenario.

Underground vs. Aboveground Shelters

One of the primary decisions is whether to build your shelter underground or aboveground. Each option has its advantages and drawbacks, and the best choice depends on your specific circumstances, including local geography, soil conditions, and budget.

Underground Shelters:

Advantages:

Superior Radiation Protection: Earth acts as an excellent shield against radiation, making underground shelters highly effective.

Thermal Insulation: Underground shelters maintain a more consistent temperature, providing better insulation from extreme weather conditions.

Concealment: Underground shelters are naturally more concealed, reducing the risk of discovery or damage.

Disadvantages:

Construction Complexity: Building underground requires excavation and potentially dealing with issues like groundwater or unstable soil, increasing costs and construction time.

Flood Risk: Underground shelters are more vulnerable to flooding, especially in areas with high water tables or heavy rainfall.

Aboveground Shelters:

Advantages:

Ease of Construction: Aboveground shelters are easier to build and modify, and they avoid the complications associated with excavation.

Accessibility: These shelters are generally easier to access, especially for individuals with mobility challenges.

Adaptability: An aboveground shelter can be integrated into existing structures or serve multiple purposes (e.g., a reinforced garage or shed).

Disadvantages:

Reduced Radiation Protection: Aboveground shelters require much thicker walls and ceilings to provide the same level of radiation protection as an underground shelter.

Exposure to Blast Effects: Aboveground shelters are more exposed to shockwaves, flying debris, and other blast-related hazards.

Soil and Geological Considerations

The type of soil and underlying geology at your chosen location can significantly affect the feasibility and effectiveness of an underground shelter.

Key Considerations:

Soil Stability: Stable, well-drained soil is ideal for underground shelters. Avoid areas with loose, sandy, or clay-heavy soils that are prone to shifting or erosion, as these can compromise the shelter's structural integrity.

Water Table: The depth of the water table is crucial. High water tables increase the risk of flooding and may require extensive waterproofing or sump pump systems. In some cases, a high water table may make underground construction impractical.

Bedrock and Obstacles: While bedrock can provide a stable foundation, it may require more effort and expense to excavate. Similarly, buried utilities, tree roots, or other obstacles can complicate construction and should be identified and planned for in advance.

Local Topography and Climate

The natural landscape and climate of your area also play a role in determining the best location for your shelter. These factors affect not only the construction process but also the long-term viability of the shelter.

Key Considerations:

Elevation: Choose a location with a slight elevation to reduce the risk of flooding. Avoid low-lying areas, riverbanks, or valleys that could become waterlogged during heavy rain or spring thaws.

Slope: A gentle slope can aid in drainage and provide natural protection from certain hazards. However, steep slopes may pose construction challenges or increase the risk of landslides.

Prevailing Winds: Consider the direction of prevailing winds in your area. Positioning your shelter upwind of potential fallout sources (e.g., a nearby city or industrial area) can reduce the risk of contamination.

Temperature Extremes: In areas with extreme temperatures, ensure your shelter is well-insulated or located underground to maintain a stable, livable environment.

Integration with Existing Structures

Retrofitting an existing structure into a fallout shelter can be a cost-effective and practical solution, especially if you have limited space or resources for a standalone shelter.

Key Considerations:

Basement Conversion: Converting a basement into a fallout shelter is one of the most common approaches. Basements are already partially underground, providing natural radiation protection. Reinforce walls and ceilings with additional shielding, and ensure the space is sealed against air and water infiltration.

Garage or Shed Conversion: Detached garages or sheds can be reinforced and converted into aboveground shelters. These structures can serve dual purposes, but they require careful planning to ensure they provide adequate protection.

Interior Rooms: If an underground or detached shelter is not feasible, consider retrofitting an interior room, such as a pantry or closet, into a fallout shelter. This option requires the addition of shielding materials and may involve reinforcing walls, doors, and windows.

Safety and Security

The shelter's location should also prioritize safety and security, both in terms of protection from external threats and ensuring the well-being of occupants.

Key Considerations:

Seismic Activity: In earthquake-prone areas, the shelter must be designed to withstand seismic events. This may involve additional structural reinforcement or selecting a location less prone to ground movement.

Access Control: Ensure that access to the shelter is secure. Reinforced doors with locks, hidden entrances, or secure tunnels can prevent unauthorized access and protect against intruders.

Proximity to Utilities: Consider the proximity to essential utilities like power, water, and communication lines. While the shelter should be self-sufficient, being near utilities can simplify construction and maintenance.

Legal and Zoning Considerations

Before building your shelter, it's essential to understand any legal and zoning regulations that may affect your project. These can vary widely depending on your location.

Key Considerations:

Permits: Check with local authorities to determine if you need permits for shelter construction, especially if it involves significant excavation or structural modifications.

Zoning Laws: Ensure your chosen location complies with local zoning laws, which may restrict certain types of construction or require specific setbacks from property lines or roads.

Homeowners' Association (HOA) Rules: If you live in a community governed by an HOA, review their rules to ensure your shelter project is allowed and complies with any design or construction standards.

Future Expansion and Flexibility

Consider whether the location allows for future expansion or modifications, as your needs or circumstances may change over time.

Key Considerations:

Space for Additions: Ensure there is enough space around the shelter for potential expansions, such as adding additional rooms, storage areas, or access points.

Modularity: Design the shelter to be modular, allowing for future upgrades or the addition of new features, such as improved ventilation systems, communication equipment, or power sources.

Landscaping: Plan landscaping around the shelter that enhances security and camouflage while allowing for potential modifications in the future.

Cost Considerations

The cost of constructing a fallout shelter can vary widely depending on location, materials, and the complexity of the design. It's essential to balance cost with the level of protection and comfort you require.

Key Considerations:

Budgeting: Establish a clear budget that includes all aspects of construction, from excavation and materials to labor and permits. Consider whether a more expensive underground location is justified by the added protection.

Cost vs. Benefit: Weigh the benefits of a more secure, concealed, or better-shielded location against the potential costs. In some cases, a simpler, less expensive shelter may be sufficient for your needs.

Phased Construction: If budget constraints are a concern, consider a phased construction approach. This allows you to build the most critical components of the shelter first, such as the core living area and basic shielding, and add additional features like enhanced ventilation, power systems, or expanded space as resources allow. This method can make the project more financially manageable while still providing essential protection in the short term.

Selecting the right location for your fallout shelter is a multifaceted decision that requires careful consideration of various factors, including proximity to your home, the type of shelter (underground or aboveground), local soil and geological conditions, and the potential hazards you are preparing for.

Excavation and Foundation: Building Underground

Building an underground fallout shelter involves significant excavation and the creation of a strong foundation to ensure the structure's stability and effectiveness. The excavation and foundation stages are critical because they set the groundwork for the entire shelter, determining its durability, safety, and protective capabilities. This chapter will guide you through the essential steps and considerations for successfully excavating and constructing the foundation for your underground shelter.

Planning the Excavation

Proper planning is essential before breaking ground on your underground shelter. Excavation is a complex process that requires careful consideration of site conditions, safety protocols, and construction methods.

Key Considerations:

Site Assessment: Before excavation, conduct a thorough site assessment. This includes testing soil stability, determining the water table level, and identifying any underground utilities or obstacles (such as tree roots or bedrock) that could interfere with excavation.

Permits and Regulations: Obtain the necessary permits and adhere to local building codes and regulations. These may dictate the depth of excavation, structural requirements, and safety measures.

Excavation Depth: The depth of your excavation will depend on the desired size and layout of your shelter. Generally, the deeper the shelter, the better the protection from radiation and blast effects. However, deeper excavations may encounter groundwater or require more extensive structural support.

Excavation Method: Choose the appropriate excavation method based on the size and depth of your shelter. For smaller projects, manual digging with shovels may suffice, while larger or deeper shelters may require heavy machinery, such as backhoes or excavators.

Excavation Techniques

The actual process of excavation involves removing soil to create the space for your shelter. This step must be done carefully to ensure the stability of the surrounding earth and to prepare the site for the foundation.

Key Techniques:

Staging the Excavation: Begin by marking the outline of your shelter on the ground. Start digging from the outer edges and work your way inward, ensuring that the walls of the excavation remain stable and do not collapse.

Shoring and Bracing: As you dig deeper, it's essential to use shoring or bracing techniques to prevent the walls of the excavation from caving in. This can involve using wooden or metal supports to reinforce the sides of the pit.

Managing Spoil: The soil removed during excavation is called "spoil." Plan for the temporary storage and eventual disposal of spoil. Some of the removed soil can be used later for backfilling or as additional shielding material.

Water Management: If groundwater is encountered during excavation, you'll need to manage it carefully to prevent flooding. This may involve using pumps to remove water or installing drainage systems around the site.

Preparing the Foundation

The foundation is the most critical structural element of your underground shelter, providing stability and support for the entire structure. A well-constructed foundation will ensure that the shelter can withstand the weight of the earth above it, as well as potential environmental stresses.

Key Considerations:

Foundation Type: The type of foundation you choose will depend on the soil conditions, the depth of the shelter, and the materials used in construction. Common foundation types for underground shelters include slab-on-grade, pier, and deep foundations (such as caissons or piles).

Levelling the Base: Once excavation is complete, the base of the pit should be leveled and compacted to create a stable surface for the foundation. This may involve adding a layer of gravel or crushed stone to improve drainage and prevent settling.

Reinforcement: Reinforce the foundation with steel rebar to increase its strength and durability. The rebar grid should be placed according to the design specifications, ensuring that it is evenly distributed across the entire foundation area.

Concrete Pouring: Pour concrete into the foundation formwork to create a solid, stable base. Ensure that the concrete is poured evenly and vibrated to remove air pockets, which can weaken the structure. Allow the concrete to cure properly before proceeding with further construction.

Waterproofing the Foundation

Waterproofing is essential for underground shelters to prevent moisture infiltration, which can lead to structural damage, mold growth, and other hazards.

Key Techniques:

Waterproof Membranes: Apply a waterproof membrane to the exterior of the foundation walls. These membranes are typically made from rubberized asphalt or other flexible materials that can withstand the pressure of the surrounding soil.

Drainage Systems: Install a drainage system around the perimeter of the foundation to divert groundwater away from the shelter. This often includes a French drain (a trench filled with gravel containing a perforated pipe) that channels water away from the foundation.

Sealants and Coatings: Use sealants or waterproof coatings on the interior surfaces of the foundation to provide an additional layer of protection against moisture. These can be applied directly to the concrete or as part of a composite system with membranes and drainage.

Structural Reinforcement and Shielding

The walls and roof of the shelter must be designed to withstand the pressure of the earth above, as well as provide adequate radiation shielding.

Key Considerations:

Wall Construction: The walls of the shelter should be constructed with reinforced concrete or another durable material capable of supporting the weight of the overlying soil. Walls should be at least 12 inches thick, depending on the level of protection required.

Roof Support: The roof of the shelter must be similarly reinforced to prevent collapse under the weight of the earth. Steel beams, rebar, and thick concrete are commonly used to create a strong, durable roof structure.

Additional Shielding: Consider adding additional layers of shielding material, such as lead sheeting, steel plates, or earth berms, to enhance the shelter's protection against radiation and blast effects. The specific materials and thicknesses will depend on your proximity to potential blast zones and the anticipated level of fallout.

Entrance and Egress

The design of the shelter's entrance is crucial for both security and practicality. It must be well-protected but also provide easy access in an emergency.

Key Considerations:

Entrance Location: The entrance should be located in a way that minimizes exposure to the outside environment. An entrance located at an angle or with a protective overhang can help reduce the amount of fallout or debris that enters the shelter.

Blast Door: Install a reinforced blast door at the entrance to protect against shockwaves, debris, and unauthorized entry. The door should be made of heavy steel or another durable material and should include secure locking mechanisms.

Emergency Exit: In addition to the main entrance, consider designing an emergency exit that can be used if the primary entrance is blocked. This could be a secondary tunnel, escape hatch, or other discreet exit point, concealed and well-protected.

Ventilation and Air Filtration

Proper ventilation is critical in an underground shelter to ensure a supply of fresh air and to remove carbon dioxide and other contaminants.

Key Considerations:

Ventilation Shafts: Include ventilation shafts that extend from the shelter to the surface. These should be equipped with blast valves and filtration systems to prevent contaminated air or debris from entering the shelter.

Air Filtration: Install a high-efficiency air filtration system, including HEPA filters and activated carbon filters, to remove radioactive particles, chemical contaminants, and other hazardous materials from the air.

Manual Ventilation: Include a manual ventilation system, such as a hand-crank blower, to ensure air circulation even if electrical power is unavailable.

Backfilling and Finalizing the Excavation

After the shelter structure is complete, the excavation must be backfilled to restore the surrounding landscape and provide additional protection.

Key Considerations:

Backfilling with Soil: Carefully backfill the excavation around the shelter with soil, ensuring that the earth is evenly distributed and compacted to prevent settling. The backfill material should be free of debris and large rocks that could damage the shelter.

Compaction: Compact the soil in layers as you backfill to prevent future settling, which can compromise the shelter's structural integrity and waterproofing.

Surface Drainage: Grade the soil around the shelter to direct water away from the structure. Surface drainage systems, such as swales or berms, can help manage runoff and prevent water from pooling around the shelter.

Final Inspections and Testing

Once construction is complete, it's essential to conduct thorough inspections and testing to ensure that the shelter is fully operational and ready for use.

Key Considerations:

Structural Integrity: Inspect the shelter's structure for any signs of weakness, cracks, or defects. Ensure that all reinforcements and supports are secure and functioning as intended.

Waterproofing Effectiveness: Test the waterproofing systems by checking for any signs of moisture or leaks in the interior of the shelter. Address any issues immediately to prevent long-term damage.

Ventilation and Air Quality: Test the ventilation and air filtration systems to ensure they are providing adequate airflow and removing contaminants effectively. Ensure that manual ventilation systems are easy to operate and functioning properly.

Access and Security: Test the entrance and exit systems, including doors, locks, and emergency exits, to ensure they are secure and easy to use in an emergency.

Excavation and foundation work are the bedrock of your underground fallout shelter, setting the stage for the entire structure's safety, durability, and effectiveness. By carefully planning the excavation, reinforcing the foundation, ensuring proper waterproofing, and integrating essential features like ventilation and secure entrances, you can create a shelter that provides reliable protection against the dangers of nuclear fallout. Proper construction practices, thorough inspections, and attention to detail will ensure that your underground shelter is not only a safe haven in times of crisis but also a durable and lasting structure for

Above-Ground Shelters: Pros and Cons

Above-ground fallout shelters offer a viable alternative to underground shelters, especially when soil conditions, water tables, or other factors make underground construction impractical. While above-ground shelters can provide effective protection from radiation and other dangers, they come with their own set of advantages and challenges. This chapter explores the pros and cons of above-ground shelters, helping you determine if this type of shelter is the right choice for your needs.

Pros of Above-Ground Shelters

Ease of Construction

Simplified Building Process: Constructing an above-ground shelter is often less complicated than building underground. It avoids the need for extensive excavation, shoring, and dealing with groundwater or unstable soil, which can make the project more straightforward and quicker to complete.

Accessibility of Construction Materials: Building materials for above-ground shelters, such as concrete, steel, and reinforced bricks, are readily available and familiar to most contractors, which can reduce costs and construction time.

Cost-Effectiveness

Lower Excavation Costs: Without the need for deep excavation, the overall cost of building an above-ground shelter can be significantly lower. This makes it a more affordable option for many people.

Reduced Water Management Requirements: Above-ground shelters are less prone to issues related to groundwater and flooding, which can further reduce construction and maintenance costs.

Adaptability and Multi-Use Potential

Dual-Purpose Structures: Above-ground shelters can be designed to serve multiple purposes. For example, a shelter could double as a fortified garage, storage area, or workshop when not in use as a fallout shelter. This dual-use approach maximizes the investment and space utilization.

Easier Integration with Existing Structures: Above-ground shelters can be more easily integrated into existing buildings or added to a home as an extension. This makes them a flexible option for homeowners looking to add a shelter without extensive renovation.

Accessibility

Better Access for All: Above-ground shelters are generally easier to access, especially for individuals with mobility issues. There are no stairs or steep inclines to navigate, making it safer and more convenient for everyone.

Emergency Access: In the event of an emergency, reaching an above-ground shelter is often quicker and less physically demanding, which can be crucial in situations where time is of the essence.

Ventilation and Air Quality

Improved Natural Ventilation: Above-ground shelters can benefit from better natural ventilation options, reducing the reliance on mechanical systems. Windows with blast shields or protected air vents can provide fresh air while maintaining protection.

Reduced Risk of Radon Exposure: Unlike underground shelters, which can be susceptible to radon gas accumulation, above-ground shelters are less likely to encounter this issue, making air quality management simpler.

Cons of Above-Ground Shelters

Reduced Radiation Shielding

Less Natural Shielding: Unlike underground shelters, which benefit from the earth's natural radiation shielding properties, above-ground shelters require much thicker walls and ceilings to achieve the same level of protection. This can make construction more challenging and costly.

Exposure to Fallout: Above-ground shelters are more directly exposed to radioactive fallout. Special attention must be paid to sealing and shielding all surfaces to prevent contamination.

Vulnerability to Blast Effects

Susceptibility to Shockwaves and Debris: Above-ground shelters are more vulnerable to blast effects from a nearby nuclear explosion, such as shockwaves and flying debris. Reinforcing the structure to withstand these forces requires additional materials and engineering, which can increase costs.

Damage from Falling Objects: Trees, power lines, and other structures may fall on or damage an above-ground shelter during a blast or other catastrophic events. The shelter must be built to withstand such impacts, which may require heavy-duty construction techniques.

Visual and Aesthetic Considerations

Aesthetic Impact: An above-ground shelter may not blend in with the surrounding environment or the design of your home, potentially affecting the aesthetics of your property. This could be a concern if you live in a neighborhood with strict homeowners' association rules or if property value is a consideration.

Lack of Concealment: Above-ground shelters are more visible and can draw unwanted attention in a post-disaster scenario, especially if resources are scarce and security becomes a concern. Camouflage or architectural integration can help, but complete concealment is difficult.

Environmental Exposure

Weather Vulnerability: Above-ground shelters are exposed to the elements, including extreme temperatures, wind, rain, and snow. This exposure can lead to higher maintenance costs and the need for robust insulation and waterproofing to ensure the shelter remains habitable.

Temperature Regulation: Maintaining a stable, comfortable temperature in an above-ground shelter can be more challenging compared to underground shelters. It may require additional heating, cooling, or insulation solutions, which can add to the complexity and cost.

Structural Requirements

Thicker Walls and Roofs: To achieve adequate radiation shielding, above-ground shelters require walls and roofs made of dense materials like reinforced concrete, which must be significantly thicker than those of underground shelters. This adds to the construction complexity and cost.

Advanced Engineering: Engineering an above-ground shelter to resist both radiation and blast effects requires careful planning and design. Structural reinforcements, such as steel beams and blast-resistant doors, may be necessary to ensure the shelter's integrity.

Above-ground fallout shelters offer a practical alternative to underground shelters, particularly in situations where excavation is impractical or too costly. They are generally easier and quicker to build, more accessible, and can often serve multiple purposes, making them an attractive option for many homeowners. However, they also come with significant challenges, including the need for enhanced radiation shielding, greater vulnerability to blast effects, and potential aesthetic and security concerns. When considering an above-ground shelter, it's crucial to weigh these pros and cons against your specific needs, budget, and location. With careful planning, proper materials, and expert engineering, an above-ground shelter can provide effective protection in the event of a nuclear disaster, while also offering the flexibility and convenience of a more accessible and multi-use structure.

Ventilation Systems: Clean Air in a Fallout Shelter

Ensuring a supply of clean, breathable air is one of the most critical aspects of fallout shelter design. In a sealed environment, air quality can quickly degrade due to the buildup of carbon dioxide, humidity, and potentially harmful contaminants. A well-designed ventilation system is essential for maintaining air quality, removing pollutants, and providing a continuous supply of fresh air. This chapter will explore the key considerations and components of effective ventilation systems in a fallout shelter, focusing on maintaining clean air during extended stays.

The Importance of Ventilation

In a confined space like a fallout shelter, the air can become stale and contaminated if not properly ventilated. Ventilation serves several crucial functions:

Oxygen Supply: Fresh air is necessary to provide oxygen for breathing and to prevent carbon dioxide from reaching dangerous levels.

Humidity Control: Ventilation helps manage humidity, reducing the risk of condensation, mold, and mildew, which can compromise both air quality and structural integrity.

Contaminant Removal: Effective ventilation systems filter out airborne contaminants, including radioactive particles, chemical pollutants, and pathogens, ensuring that the air remains safe to breathe.

Types of Ventilation Systems

There are several types of ventilation systems that can be used in a fallout shelter, each with its own advantages and limitations. The choice of system depends on factors such as shelter size, expected occupancy, and available power sources.

Natural Ventilation:

Pros: Simple and requires no power; can be integrated into the shelter's design through strategically placed vents and openings.

Cons: Limited control over airflow; may not provide adequate filtration or protection from airborne contaminants.

Best Used: In less critical situations or as a supplementary system to powered ventilation.

Powered Ventilation:

Pros: Provides controlled airflow and can be equipped with advanced filtration systems; capable of maintaining a stable air supply and removing contaminants efficiently.

Cons: Requires a reliable power source; more complex and costly to install and maintain.

Best Used: In larger shelters or those intended for long-term occupancy where maintaining air quality is critical.

Manual Ventilation:

Pros: Operated by hand, making it independent of electricity; can be used as a backup system if power fails.

Cons: Requires manual effort, which may be difficult over extended periods; limited in capacity and airflow compared to powered systems.

Best Used: As an emergency backup or in conjunction with a primary powered ventilation system.

Components of an Effective Ventilation System

A comprehensive ventilation system in a fallout shelter consists of several key components, each designed to ensure that air is circulated, filtered, and safe for occupants.

Air Intake and Exhaust Vents:

Air Intake: Fresh air enters the shelter through an air intake vent, which should be placed away from potential contamination sources. The intake should be equipped with blast valves or blast-resistant filters to protect against shockwaves and debris.

Air Exhaust: Stale air exits the shelter through an exhaust vent, which should be positioned to encourage effective air circulation. Exhaust vents may also require blast protection to ensure they remain functional during a nuclear event.

Air Filters:

HEPA Filters: High-Efficiency Particulate Air (HEPA) filters are essential for removing radioactive particles, dust, and other fine contaminants from the air. These filters can capture particles as small as 0.3 microns with a high degree of efficiency.

Activated Carbon Filters: Activated carbon filters are used to remove chemical vapors, odors, and certain gases from the air. They work by adsorbing pollutants onto the surface of the carbon particles.

Pre-Filters: Pre-filters are installed before HEPA or carbon filters to capture larger particles, extending the life of the main filters and improving overall system efficiency.

Fans and Blowers:

Powered Fans: Powered fans are used to move air through the ventilation system, ensuring a continuous supply of fresh air and the removal of stale air. These fans can be controlled to regulate airflow based on occupancy and conditions inside the shelter.

Manual Blowers: Manual blowers, such as hand-crank ventilators, can be used in case of a power outage. While less efficient than powered fans, they provide an essential backup to keep air circulating.

Ductwork:

Air Ducts: Ductwork channels the air throughout the shelter, connecting intake and exhaust vents with the living spaces. Ducts should be designed to minimize resistance and ensure even distribution of air.

Blast Dampers: Ducts may need to be equipped with blast dampers, which close automatically in the event of an explosion, preventing shockwaves from entering the shelter.

Air Quality Monitors:

CO2 Monitors: Carbon dioxide monitors track the levels of CO2 inside the shelter, providing alerts if concentrations approach unsafe levels. This helps ensure that the ventilation system is providing sufficient fresh air.

Radon and Other Gas Detectors: Depending on the shelter's location, it may be necessary to monitor for radon or other harmful gases that could seep into the shelter.

Designing an Efficient Ventilation System

Designing an effective ventilation system requires careful planning and consideration of various factors, including the shelter's size, occupancy, and environmental conditions.

Calculating Airflow Requirements:

Occupancy: Determine the maximum number of occupants and calculate the total volume of the shelter. The ventilation system must be capable of supplying enough fresh air to maintain a safe and comfortable environment.

Air Changes Per Hour (ACH): The ACH is a measure of how many times the air in the shelter is replaced with fresh air each hour. For fallout shelters, a typical ACH of 6-10 is recommended, meaning the entire volume of air should be replaced 6-10 times per hour.

Ventilation System Layout:

Intake and Exhaust Placement: Position air intake vents at the lowest possible point, where cool, fresh air can enter. Exhaust vents should be placed at the highest point, where warm, stale air naturally rises. This placement takes advantage of natural convection to aid airflow.

Duct Design: Design the ductwork to minimize bends and obstructions, which can reduce airflow efficiency. Use smooth, rigid ducts where possible, as flexible ducts can create turbulence and increase resistance.

Filtration and Maintenance:

Filter Maintenance: Regularly inspect and replace air filters according to the manufacturer's recommendations. A clogged filter can significantly reduce airflow and compromise air quality.

Access Panels: Install access panels for easy maintenance and filter replacement. This ensures that the system remains operational and efficient over time.

Backup and Redundancy:

Power Backup: Consider a backup power source, such as a generator or battery system, to keep powered ventilation running during an extended power outage. Alternatively, manual blowers should be included as a fail-safe.

Redundant Systems: In critical shelters, having redundant ventilation systems can provide a safety net in case the primary system fails. This might include a second set of intake and exhaust vents or an independent manual system.

Managing Air Contaminants

In a fallout shelter, air contaminants can come from external sources, such as radioactive fallout, or internal sources, like carbon dioxide and moisture from occupants.

Radioactive Fallout:

Filtering Radioactive Particles: Use HEPA filters to remove radioactive particles from the incoming air. Regularly monitor the outside air quality, especially after a nuclear event, to adjust ventilation settings as needed.

Sealing Air Leaks: Ensure that all vents, doors, and windows are properly sealed to prevent unfiltered air from entering the shelter.

Carbon Dioxide (CO2):

CO2 Scrubbers: In addition to ventilation, consider installing CO2 scrubbers, which chemically remove carbon dioxide from the air. This is particularly important in situations where airflow is limited or when occupants must stay sealed inside the shelter for extended periods.

Ventilation Rate: Ensure that the ventilation rate is sufficient to keep CO2 levels below 1,000 parts per million (ppm), the threshold generally considered safe for prolonged exposure.

Humidity Control:

Dehumidifiers: Use dehumidifiers to control moisture levels inside the shelter. High humidity can lead to mold growth, which can negatively affect air quality and health.

Ventilation: Proper ventilation also helps to manage humidity by removing moisture-laden air and bringing in drier air from outside.

Chemical and Biological Contaminants:

Activated Carbon Filters: These filters are essential for removing chemical vapors and odors. They should be replaced regularly to maintain effectiveness.

UV Air Purifiers: Consider adding UV air purifiers to the system to kill airborne bacteria and viruses, adding an extra layer of protection.

Testing and Maintenance

Once the ventilation system is installed, regular testing and maintenance are crucial to ensure it functions correctly during an emergency.

System Testing:

Flow Testing: Perform airflow testing to ensure that the system meets the required ACH for the shelter's size and occupancy.

Filter Efficiency: Test the filtration system periodically to ensure it effectively removes contaminants. Replace filters as needed.

Routine Maintenance:

Inspection Schedule: Establish a regular inspection schedule for the ventilation system. This should include checking for leaks, testing airflow, and inspecting filter conditions.

Emergency Drills: Conduct emergency drills to practice operating manual ventilation systems and switching to backup power if necessary.

A well-designed ventilation system is essential for ensuring the safety and comfort of a fallout shelter's occupants. Effective ventilation not only ensures a continuous supply of oxygen but also protects against the buildup of harmful gases, humidity, and airborne contaminants, making your fallout shelter a secure and liveable space during an extended stay.

Long-Term Considerations

In scenarios where sheltering could last for weeks or even months, it's essential to plan for the long-term functionality of your ventilation system. This includes ensuring that the system can operate independently if external power sources are compromised and that it can be maintained over an extended period without outside assistance.

Key Considerations:

Extended Filter Life: For long-term sheltering, consider stocking multiple replacement filters, especially HEPA and activated carbon filters. Estimate how often these will need to be replaced based on your ventilation rate and potential contamination levels.

Energy Efficiency: If your ventilation system relies on electrical power, prioritize energy-efficient components to minimize the load on your power supply, whether it's a generator, solar panels, or batteries. Manual ventilation systems should be ergonomic and designed for ease of use over extended periods.

Continuous Monitoring: Equip your shelter with sensors that provide real-time data on air quality, CO_2 levels, and humidity. These sensors can alert you to any changes in air quality that might require adjustments to ventilation settings or maintenance tasks.

Redundancy and Backup Systems:

Secondary Air Supply: Consider a secondary air intake, which can be used if the primary intake becomes compromised. This could be as simple as a secondary vent with a manual override or a completely separate intake system.

Manual Operation: Ensure that all critical functions of the ventilation system, including fan operation and filter changes, can be performed manually in case of a power failure. Having a manual backup is crucial for long-term survival in a post-fallout scenario.

Ventilation System Durability:

High-Quality Materials: Use durable materials for ducts, fans, and other components to withstand the potentially harsh conditions inside a shelter over time. This reduces the likelihood of system failure due to wear and tear.

Regular Drills and Training: Conduct regular drills to ensure all occupants are familiar with the operation of the ventilation system, including how to switch between power modes, replace filters, and operate manual blowers. Training everyone in these procedures ensures that the system can be maintained effectively even if the primary operator is incapacitated.

Enhancing Shelter Liveability with Advanced Ventilation Features

While basic ventilation systems are sufficient to keep air clean and breathable, advanced features can significantly enhance the liveability of your fallout shelter, particularly during long-term occupancy.

Climate Control Integration:

Air Conditioning and Heating: Integrating climate control into your ventilation system can help maintain a comfortable temperature inside the shelter. This is especially important in areas with extreme weather conditions.

Humidity Control: Advanced humidity control systems, including dehumidifiers or humidifiers integrated into the ventilation system, can ensure that the air remains within a comfortable and safe humidity range.

Air Quality Enhancements:

Air Purification Systems: In addition to standard filtration, consider installing air purifiers that use UV light, ionization, or other technologies to eliminate airborne pathogens, allergens, and even some chemical contaminants.

Aromatherapy: While not essential for survival, incorporating aromatherapy diffusers into the ventilation system can improve mental well-being and comfort by releasing calming scents into the air.

Noise Reduction:

Silent Operation: Noise from ventilation systems can become a source of stress, especially over long periods. Choose components designed for silent operation, or install noise-reducing features such as soundproof ducts or vibration dampeners to minimize disturbances.

Smart Controls:

Automated Systems: Automating your ventilation system with smart controls can make it easier to manage air quality without constant manual adjustments. Smart sensors can adjust fan speed, airflow, and filter use based on real-time conditions, optimizing efficiency and comfort.

Remote Monitoring: If your shelter is part of a larger preparedness plan that includes remote surveillance, consider integrating remote monitoring of the ventilation system. This allows you to check on air quality and system status even when you're not in the shelter.

Final Preparations and Checklist

Before considering your shelter's ventilation system complete, it's important to go through a final checklist to ensure everything is in place and functioning correctly.

Final Checklist:

System Testing: Test all components of the ventilation system under normal conditions and simulate an emergency to ensure it performs as expected. Check airflow, filter efficiency, and the functionality of backup systems.

Supplies Stocked: Verify that you have a sufficient stockpile of filters, parts, and tools for maintaining the ventilation system over an extended period.

Manuals and Instructions: Keep detailed manuals and instructions on hand for all system components, including troubleshooting guides for common issues. Ensure that all occupants know where these documents are located and understand the basics of system operation.

Emergency Drills: Conduct a full emergency drill to practice switching to manual ventilation, changing filters, and responding to changes in air quality. Include all shelter occupants in the drill to ensure everyone is prepared.

A reliable and efficient ventilation system is the cornerstone of a safe and habitable fallout shelter. Proper installation, regular maintenance, and ongoing monitoring of the system will help you maintain clean air and protect the health and well-being of everyone in the shelter. Whether for short-term use or extended stays, a well-designed ventilation system is essential for survival and comfort in a fallout shelter.

Filtration and Purification: Ensuring Safe Air Supply

In a fallout shelter, ensuring a safe and clean air supply is crucial for the health and survival of its occupants. Filtration and purification systems are designed to remove harmful particles, gases, and other contaminants from the air, making it safe to breathe even in a heavily contaminated environment. This chapter will explore the various methods and technologies available for air filtration and purification, providing guidance on how to implement these systems effectively in your shelter.

The Importance of Filtration and Purification

In the event of nuclear fallout, the air outside your shelter may contain radioactive particles, toxic chemicals, and biological contaminants. Without proper filtration and purification, these hazardous materials could enter your shelter and pose serious health risks. The primary goals of an air filtration and purification system are to:

Remove Radioactive Particles: These include alpha, beta, and gamma-emitting particles that can cause radiation sickness and increase the risk of cancer.

Eliminate Toxic Chemicals: Fallout may include chemical pollutants from fires, industrial accidents, or the explosion itself. These chemicals can cause respiratory issues and other health problems.

Filter out Biological Contaminants: Bacteria, viruses, and spores can be present in the air after a disaster, posing a risk of infection and disease.

Control Humidity and Air Quality: In addition to removing harmful substances, filtration systems help maintain a healthy indoor environment by controlling humidity and reducing allergens.

Types of Airborne Contaminants

Understanding the types of contaminants you may encounter helps in selecting the appropriate filtration and purification methods. The primary airborne contaminants include:

Radioactive Particles:

Alpha Particles: These are relatively large and can be filtered out effectively by most filtration systems. However, they are highly dangerous if inhaled or ingested.

Beta Particles: Smaller than alpha particles, beta particles can penetrate certain materials but are still relatively easy to filter out with appropriate systems.

Gamma Rays: These are highly penetrating and require dense materials like lead or thick concrete to block. However, gamma rays are less of a concern for filtration as they do not behave like particles that can be trapped in filters.

Chemical Contaminants:

Volatile Organic Compounds (VOCs): These chemicals can evaporate into the air from substances like paint, fuel, and industrial chemicals. They are often removed using activated carbon filters.

Toxic Gases: These may include carbon monoxide, chlorine, and other hazardous gases that can enter the shelter if not properly filtered.

Biological Contaminants:

Bacteria and Viruses: These can spread through the air and are typically removed by HEPA filters or neutralized by ultraviolet (UV) light.

Mold Spores: Mold can grow in humid environments and release spores into the air, which can cause respiratory issues. Proper filtration and humidity control help mitigate this risk.

Filtration Technologies

Several filtration technologies are available to address different types of contaminants. Understanding these technologies is key to designing an effective air filtration system for your shelter.

HEPA Filters (High-Efficiency Particulate Air):

Function: HEPA filters are designed to capture particles as small as 0.3 microns with an efficiency of 99.97%. This makes them ideal for removing radioactive particles, dust, and other fine contaminants.

Application: HEPA filters are commonly used in air purifiers and ventilation systems. They are essential in any fallout shelter to ensure that airborne radioactive particles are removed from the incoming air.

Maintenance: Regular replacement of HEPA filters is necessary to maintain their effectiveness. Check manufacturer recommendations for replacement intervals based on usage and contamination levels.

Activated Carbon Filters:

Function: Activated carbon filters are highly effective at removing chemical contaminants, odors, and VOCs. They work by adsorbing these substances onto the surface of the carbon granules.

Application: These filters are often used in conjunction with HEPA filters in air purification systems to provide comprehensive protection against both particulate and chemical contaminants.

Maintenance: Like HEPA filters, activated carbon filters need regular replacement, especially in environments with high levels of chemical pollutants.

Pre-Filters:

Function: Pre-filters are used to capture larger particles such as dust, hair, and debris before the air reaches the HEPA or activated carbon filters. This extends the life of the more expensive filters and improves overall system efficiency.

Application: Pre-filters are typically the first stage in a multi-layer filtration system.

Maintenance: Pre-filters should be cleaned or replaced regularly to prevent clogging and maintain airflow.

UV-C Light (Ultraviolet Germicidal Irradiation):

Function: UV-C light kills or inactivates bacteria, viruses, and other microorganisms by damaging their DNA. This method is particularly effective for neutralizing airborne biological contaminants.

Application: UV-C light can be integrated into air filtration systems as a secondary measure after physical filtration. It is especially useful in environments where there is a high risk of biological contamination.

Maintenance: UV bulbs need to be replaced periodically as their effectiveness diminishes over time. Ensure that the UV light is properly shielded to prevent exposure to shelter occupants.

Electrostatic Filters:

Function: Electrostatic filters use an electric charge to capture particles in the air. As air passes through, particles are attracted to oppositely charged plates or fibers.

Application: These filters are sometimes used in combination with other filtration methods to enhance particle removal. However, they are less effective at capturing smaller particles compared to HEPA filters.

Maintenance: Electrostatic filters require regular cleaning to maintain their effectiveness. The collection plates must be wiped down or washed to remove accumulated particles.

Purification Technologies

In addition to filtration, air purification technologies can help ensure that the air inside your shelter remains safe and clean. These technologies are particularly useful for removing gases and biological contaminants that filtration alone might not fully address.

Ionization and Ozone Generators:

Ionization: Ionizers release charged ions into the air that attach to particles, causing them to clump together and fall out of the air. While this can reduce airborne particulate levels, ionization does not filter the air and can produce ozone as a byproduct.

Ozone Generators: These devices produce ozone, which can neutralize odors and kill certain bacteria and viruses. However, ozone is a lung irritant and should be used cautiously, especially in occupied spaces.

Application: These technologies are generally not recommended as the primary method of air purification in a fallout shelter due to the potential health risks associated with ozone.

Photo electrochemical Oxidation (PECO):

Function: PECO technology uses light to catalyze a reaction that breaks down organic molecules, including bacteria, viruses, and VOCs, into harmless substances like carbon dioxide and water.

Application: PECO systems can be integrated into air purifiers as an advanced form of air purification, particularly effective against gaseous pollutants and pathogens.

Maintenance: PECO systems require periodic replacement of the catalyst and light source to maintain efficiency.

Designing a Filtration and Purification System

Designing an effective filtration and purification system for your fallout shelter requires careful planning and consideration of the specific threats you may face. The system should be capable of handling the shelter's air supply needs while ensuring that all contaminants are removed or neutralized.

System Layout:

Multi-Stage Filtration: Implement a multi-stage filtration system that includes pre-filters, HEPA filters, and activated carbon filters. This ensures that all types of contaminants are addressed, from large particles to chemical vapours.

Integration with Ventilation: The filtration system should be integrated into the shelter's ventilation system, ensuring that all incoming air is filtered before entering the living space.

Redundancy: Consider including redundant filtration systems to provide backup in case one system fails. This could include a secondary air intake with its own filtration system or portable air purifiers that can be deployed as needed.

Air Flow Management:

Positive Pressure System: Consider using a positive pressure system that pushes filtered air into the shelter, creating a slight overpressure that prevents unfiltered air from seeping in through cracks or gaps.

Air Distribution: Ensure that filtered air is evenly distributed throughout the shelter. This may involve the use of ductwork and fans to direct airflow to all areas.

Power Considerations:

Backup Power: The filtration and purification systems should be connected to a reliable power source, such as a generator or battery backup, to ensure continuous operation during an emergency.

Manual Operation: Include a manual option for air filtration, such as hand-cranked fans or bellows, to ensure that air can still be filtered if electrical power is unavailable.

Maintenance and Testing

Regular maintenance and testing are essential to ensure that your filtration and purification systems remain effective over time. Without proper upkeep, filters can become clogged, and purification systems can lose their effectiveness, compromising the safety of your shelter's air supply.

Maintenance Schedule:

Filter Replacement: Follow the manufacturer's guidelines for replacing HEPA, activated carbon, and pre-filters. Keep a stockpile of replacement filters on hand, especially if long-term sheltering is anticipated.

UV-C Bulb Replacement: Replace UV-C bulbs as recommended by the manufacturer to ensure they continue to effectively neutralize biological contaminants.

System Inspection: Regularly inspect the filtration and purification systems for signs of wear, damage, or clogging. Clean or repair components as needed to maintain optimal performance.

Air Quality Testing:

Routine Testing: Conduct regular air quality tests to monitor levels of particulate matter, carbon dioxide, humidity, and any specific contaminants relevant to your shelter's environment. Testing the air quality helps you identify any issues with your filtration and purification systems before they become serious problems.

Key Testing Parameters:

Particulate Matter (PM): Use a particle counter to measure the concentration of particulate matter in the air, particularly focusing on particles in the PM2.5 and PM10 ranges, which are the most harmful to health.

Chemical Contaminants: Test for the presence of VOCs, carbon monoxide, and other harmful gases using appropriate sensors and detectors. Ensure that the levels remain below established safety thresholds.

Radiation Levels: Regularly monitor radiation levels inside the shelter using a Geiger counter or similar device to ensure that your filtration system is effectively removing radioactive particles.

Biological Contaminants: If your shelter is at risk for biological contamination, consider using air sampling methods to detect bacteria, viruses, or mold spores. This can be particularly important if anyone in the shelter becomes ill.

Emergency Drills:

Simulate Power Failure: Conduct drills to simulate a power failure, ensuring that manual ventilation and filtration systems can be activated quickly and effectively. This practice will prepare all shelter occupants to respond appropriately in an emergency.

Filter Change Practice: Practice changing air filters under simulated emergency conditions, so everyone knows how to replace them quickly and correctly. This drill should include checking for proper filter placement and sealing.

Addressing Common Filtration and Purification Challenges

Even the best-designed filtration and purification systems can face challenges, especially during long-term use. Understanding these challenges and having solutions ready can help you maintain air quality in your shelter.

Clogged Filters:

Problem: Over time, filters can become clogged with dust, debris, and contaminants, reducing airflow and filtration efficiency.

Solution: Implement a regular maintenance schedule to check and replace filters before they become too clogged. Keep spare filters readily available and consider using pre-filters to extend the life of more expensive HEPA and carbon filters.

Power Outages:

Problem: A power outage can disable powered filtration systems, leaving the shelter without adequate air purification.

Solution: Ensure you have backup power solutions like batteries or generators. Include manual filtration options, such as hand-cranked fans or portable air purifiers, to maintain airflow and filtration during a power failure.

Filter Saturation:

Problem: Activated carbon filters can become saturated with chemicals and lose their effectiveness over time, particularly in environments with high levels of VOCs or toxic gases.

Solution: Monitor air quality for signs of filter saturation, such as persistent odors or increasing VOC levels. Replace activated carbon filters regularly and keep track of their usage life.

Maintenance Fatigue:

Problem: Long-term sheltering may lead to maintenance fatigue, where regular upkeep of the filtration system is neglected due to physical or mental exhaustion.

Solution: Rotate maintenance responsibilities among shelter occupants to distribute the workload. Establish a clear maintenance schedule and adhere to it strictly, with reminders and checklists to ensure nothing is overlooked.

Enhancing Shelter Safety with Advanced Filtration Systems

For those seeking the highest level of protection, advanced filtration and purification systems can provide additional safety measures, especially in shelters designed for long-term occupancy or in environments with severe contamination risks.

Advanced Filtration Technologies:

PECO (Photo Electrochemical Oxidation): Consider integrating PECO technology, which uses light-activated catalytic processes to break down organic molecules, including VOCs, bacteria, and viruses, into harmless by-products.

Multi-Stage Filtration Systems: Use multi-stage systems that combine pre-filters, HEPA filters, activated carbon filters, and UV-C light to provide comprehensive air purification. This approach ensures that all potential contaminants are addressed, enhancing overall air quality.

Smart Filtration Systems:

Automated Control Systems: Install smart control systems that automatically adjust fan speed, filter usage, and airflow based on real-time air quality data. These systems can optimize the efficiency of your filtration setup and ensure continuous protection without manual intervention.

Remote Monitoring and Alerts: Set up remote monitoring and alert systems that notify you of any issues with air quality, filter performance, or system malfunctions. This can be especially useful if the shelter is part of a larger emergency preparedness plan that includes remote access.

Redundant Systems:

Secondary Filtration Pathways: Create redundant filtration pathways that can be activated if the primary system fails. This might include additional air intakes with separate filtration units or portable air purifiers that can be deployed as needed.

Layered Defense: Consider a layered defense approach, where different filtration methods (e.g., mechanical filters, chemical adsorption, UV sterilization) are used in tandem to provide overlapping protection against a wide range of contaminants.

Final Preparations and Readiness

Before your filtration and purification system is considered ready, it's essential to conduct a final series of checks and preparations to ensure everything is functioning correctly and efficiently.

Final Checklist:

Comprehensive Testing: Perform a final round of air quality tests, checking for particulate levels, chemical contaminants, and overall system performance. Make any necessary adjustments or repairs before considering the system ready.

Supply Inventory: Ensure you have a full inventory of replacement filters, parts, and maintenance tools. Store these supplies in a location that is easily accessible within the shelter.

Documentation and Training: Provide detailed documentation on the operation and maintenance of the filtration system, including manuals, troubleshooting guides, and maintenance schedules. Ensure all shelter occupants are trained on how to operate and maintain the system.

Emergency Drills: Conduct a full emergency drill, simulating scenarios such as filter changes, power outages, and manual system activation. This will ensure that everyone is prepared to respond quickly and effectively in a real emergency.

A well-designed filtration and purification system is essential for maintaining a safe and breathable air supply in a fallout shelter. By understanding the various filtration technologies, implementing a robust system design, and adhering to regular maintenance and testing, you can ensure that your shelter remains a secure environment even in the most challenging conditions. Advanced filtration options and smart controls can further enhance safety, providing peace of mind that your shelter is equipped to protect against all airborne threats. Proper preparation, vigilance, and regular practice are key to ensuring that your shelter's air supply remains clean and safe for as long as necessary.

Radiation Detection: Tools and Techniques

Radiation detection is a critical component of safety in a fallout shelter. Understanding how to measure and monitor radiation levels both inside and outside the shelter is essential for making informed decisions about when it is safe to exit the shelter, as well as for ensuring that the shelter environment remains safe over time. This chapter will explore the tools and techniques used for radiation detection, providing a comprehensive guide to understanding and using these technologies effectively.

Understanding Radiation and Its Types

Before diving into the tools and techniques for radiation detection, it's important to have a basic understanding of what radiation is and the types of radiation that might be encountered in a fallout situation.

Types of Ionizing Radiation:

Alpha Particles: Alpha particles are large, positively charged particles that can be blocked by a sheet of paper or even the outer layer of human skin. However, they pose a serious health risk if inhaled or ingested.

Beta Particles: Beta particles are smaller, negatively charged particles that can penetrate the skin but are usually stopped by materials like plastic or a few millimeters of aluminum.

Gamma Rays: Gamma rays are high-energy electromagnetic waves that can penetrate most materials, including the human body. Dense materials like lead or thick concrete are required to block gamma rays effectively.

Neutrons: Neutron radiation is less common but can be encountered in nuclear explosions. Neutrons are uncharged particles that can penetrate most materials and require substances like water, polyethylene, or concrete to block them.

Tools for Radiation Detection

There are several types of devices available for detecting and measuring radiation. Each tool has specific uses and is designed to measure different aspects of radiation exposure.

Geiger-Müller Counter (Geiger Counter):

Function: A Geiger counter is a widely used radiation detection device that measures ionizing radiation, including alpha, beta, and gamma radiation. It produces an audible click or a digital readout that indicates the presence of radiation.

Application: Geiger counters are useful for general radiation detection and for checking radiation levels in and around the shelter. They are portable, easy to use, and provide immediate feedback.

Limitations: While Geiger counters are good for detecting the presence of radiation, they are less effective at measuring the exact energy or type of radiation.

Dosimeter:

Function: A dosimeter measures an individual's cumulative exposure to radiation over time. It is typically worn on the body and records the total dose of radiation absorbed.

Application: Dosimeters are essential for monitoring the radiation exposure of shelter occupants. They help ensure that individuals do not exceed safe exposure limits over the course of their stay in the shelter.

Limitations: Dosimeters provide cumulative data rather than real-time readings, so they are best used in conjunction with other detection tools.

Scintillation Detector:

Function: Scintillation detectors measure radiation by detecting the flashes of light (scintillations) produced when ionizing radiation interacts with a scintillating material, such as sodium iodide.

Application: These detectors are more sensitive than Geiger counters and can measure lower levels of radiation, making them useful for detecting gamma rays and other low-intensity radiation.

Limitations: Scintillation detectors are generally more expensive and complex than Geiger counters, requiring calibration and a more detailed understanding of their operation.

Ionization Chamber:

Function: Ionization chambers measure radiation by collecting ionized particles created by radiation in a gas-filled chamber. They are particularly useful for measuring high levels of gamma and X-ray radiation.

Application: Ionization chambers are often used in environments where precise measurement of radiation intensity is required, such as in medical settings or radiation labs.

Limitations: These devices are generally bulkier and less portable than other radiation detectors, making them less practical for personal use in a fallout shelter.

Personal Radiation Detectors (PRDs):

Function: PRDs are small, portable devices that provide real-time monitoring of radiation levels, usually with a visual or audible alarm when radiation exceeds a predefined threshold.

Application: PRDs are useful for continuous monitoring of radiation levels by individuals, providing an early warning if radiation levels suddenly increase.

Limitations: PRDs are typically limited in their ability to differentiate between types of radiation, focusing primarily on gamma and X-ray detection.

Techniques for Effective Radiation Monitoring

Using radiation detection tools effectively involves more than just turning on a device and taking readings. Proper technique ensures accurate measurements and helps you make informed decisions based on the data collected.

Baseline Measurements:

Establishing a Baseline: Before a fallout event, establish baseline radiation levels inside and outside the shelter. This helps you understand what is normal for your environment and detect any significant changes in radiation levels after the event.

Regular Monitoring: Once fallout has occurred, regularly monitor radiation levels at consistent intervals. This helps track changes over time and determines when it might be safe to leave the shelter.

Monitoring Inside the Shelter:

Air Quality Monitoring: Use a Geiger counter or other detectors to check for radiation in the air, particularly near ventilation intakes. This ensures that your filtration system is effectively removing radioactive particles.

Surface Contamination: Periodically check surfaces inside the shelter, such as walls, floors, and stored supplies, for any signs of radiation. This can help identify any breaches in the shelter's integrity or issues with contamination.

Dosimeter Usage: Ensure that all shelter occupants wear dosimeters to track individual radiation exposure. Check dosimeter readings regularly to monitor cumulative exposure levels.

Monitoring Outside the Shelter:

External Radiation Levels: If it is safe to do so, periodically check radiation levels outside the shelter. This can help you gauge when radiation levels are decreasing and assess whether it is safe to exit the shelter.

Entry Points: Pay special attention to monitoring radiation levels near entry and exit points, as these areas are more likely to experience contamination from outside.

Using PRDs and Alarms:

Continuous Monitoring: For continuous real-time monitoring, PRDs can be worn by individuals or placed in key areas around the shelter. Set alarm thresholds to provide early warnings of rising radiation levels.

Alarm Response: Establish protocols for responding to PRD alarms, including moving to safer areas within the shelter or adjusting ventilation and filtration systems.

Data Logging and Analysis:

Recording Data: Keep a detailed log of all radiation measurements, including time, location, and device readings. This information can be invaluable for tracking radiation trends and making informed decisions.

Trend Analysis: Analyze the data to identify trends in radiation levels. Decreasing levels may indicate that it is becoming safer to leave the shelter, while increasing levels could suggest a new contamination event or a problem with the shelter's integrity.

Safety Protocols When Using Radiation Detection Tools

Handling radiation detection tools requires certain safety protocols to ensure accurate readings and to protect users from potential exposure.

Proper Handling:

Device Calibration: Regularly calibrate your radiation detection tools according to the manufacturer's instructions to ensure they provide accurate readings.

Avoiding Contamination: When using detectors outside the shelter, avoid touching the device to potentially contaminated surfaces. After use, clean the device with appropriate materials to prevent cross-contamination.

Personal Protection:

Protective Gear: When monitoring radiation levels outside the shelter or in potentially contaminated areas, wear protective clothing, gloves, and a mask to minimize exposure to radioactive particles.

Distance and Shielding: Use distance and shielding to your advantage when taking readings in high-radiation areas. Stand as far away from the source as possible, and use available barriers to reduce exposure.

Device Maintenance:

Battery Check: Regularly check the batteries in your radiation detection devices and keep spare batteries on hand. Ensure that devices are fully operational at all times.

Storage: Store radiation detection tools in a safe, dry, and radiation-free environment when not in use. Protect them from physical damage and environmental factors that could impair their function.

Interpreting Radiation Readings

Understanding and interpreting the readings from your radiation detection tools is crucial for making informed decisions. Here are some key concepts to keep in mind:

Radiation Units:

Roentgens (R): Measures exposure to gamma and X-rays in the air.

Rads: Measures absorbed dose of radiation in any material.

Rems: Measures the biological effect of radiation on human tissue (used to assess risk to human health).

Sieverts (Sv): An international unit equivalent to rems, often used to measure dose equivalent, considering the type of radiation and its effect on tissues.

Thresholds and Safe Levels:

Background Radiation: Understand the normal background radiation levels for your area, which typically range from 0.05 to 0.2 micro Sieverts per hour (μSv/h).

Action Levels: Establish action levels based on guidelines from health authorities. For example, evacuation is often recommended if radiation levels exceed 10 millisieverts (mSv) per hour.

Cumulative Dose: Monitor cumulative radiation doses, especially using dosimeters. The generally accepted safe limit for emergency exposure is 50 rems (0.5 Sv) over a short period, but lower levels should be targeted to minimize long-term risk.

Making Decisions Based on Readings:

Shelter Stay Duration: Use radiation readings to determine how long to stay in the shelter. If outside radiation levels are high, it may be necessary to remain inside for an extended period.

Exit Strategy: Plan an exit strategy based on decreasing radiation levels outside. Generally, a level of 0.1 rems per hour (1 millisievert per hour) is considered relatively safe for short-term exposure, but lower levels are preferable for extended time outside the shelter.

Interpreting Anomalies:

Spikes in Radiation: Occasional spikes in radiation levels might indicate a new fallout event or the disturbance of settled radioactive particles. If detected, increase your monitoring frequency and take additional protective measures.

Decreasing Levels: A steady decrease in radiation levels over time generally indicates that radioactive materials are decaying or dispersing. This trend may suggest it's becoming safer to leave the shelter, but always wait until levels stabilize at a safe threshold before considering an exit.

Preparing for Post-Shelter Monitoring

Even after leaving the shelter, radiation monitoring remains crucial, as contaminated areas may still pose risks. Preparing for ongoing radiation detection helps ensure safety during the transition back to normal activities.

Portable Monitoring:

Field Radiation Detectors: Equip yourself with portable radiation detectors, such as handheld Geiger counters or PRDs, to monitor radiation levels as you move around outside the shelter.

Surveying Contaminated Areas: Before reentering your home or other buildings, survey these areas for radiation hotspots. Pay particular attention to places where fallout may have accumulated, such as gutters, roofs, and open ground.

Decontamination Protocols:

Personal Decontamination: After monitoring, always perform a personal decontamination routine. This includes removing and disposing of contaminated clothing, washing exposed skin, and thoroughly cleaning any equipment that was used outside.

Site Decontamination: If radiation hotspots are detected around your property, take steps to decontaminate these areas. This may involve removing contaminated soil, washing surfaces with soap and water, and sealing off any areas that remain highly radioactive.

Long-Term Monitoring:

Ongoing Surveillance: Continue to monitor radiation levels regularly, especially in areas where contamination was detected. Over time, radiation should decrease, but it's essential to remain vigilant.

Health Monitoring: Keep track of any health symptoms that may indicate radiation exposure, such as skin burns, fatigue, or unusual illnesses. Seek medical attention immediately if any concerning symptoms arise.

Radiation detection is an indispensable aspect of fallout shelter safety. By understanding the different types of radiation, using the appropriate detection tools, and following effective monitoring techniques, you can ensure that your shelter remains a safe haven during a nuclear event. Regular maintenance, proper interpretation of readings, and ongoing vigilance after leaving the shelter are all critical for protecting yourself and your loved ones from the dangers of radiation.

Equipped with the right knowledge and tools, you can confidently navigate the challenges of radiation monitoring, making informed decisions that prioritize safety and well-being. Whether you're inside the shelter or assessing conditions post-event, the ability to detect and respond to radiation effectively is key to surviving and thriving in the aftermath of a nuclear fallout.

Communication Systems for Fallout Shelters

Effective communication is crucial for survival in a fallout shelter, enabling you to stay informed about the external environment, coordinate with others, and seek help if needed. In a crisis situation, traditional communication networks may be disrupted, so it's essential to have reliable, alternative communication systems in place. This chapter will explore the various communication options available for fallout shelters, offering guidance on selecting and using these systems to maintain contact with the outside world and within the shelter.

The Importance of Communication in a Fallout Shelter

Communication serves multiple vital functions in a fallout shelter:

Receiving Information: Staying informed about the external environment, including radiation levels, weather conditions, and government directives, is essential for making safe decisions about when to leave the shelter.

Coordinating with Others: Whether communicating with family members, other shelter occupants, or external contacts, effective communication is key to coordinating actions, sharing resources, and ensuring everyone's safety.

Emergency Assistance: In the event of a medical emergency or if the shelter is compromised, communication systems can be used to call for help or coordinate a rescue.

Types of Communication Systems

Several communication systems can be used in a fallout shelter, each with its own advantages and limitations. Understanding these options will help you choose the most appropriate systems for your needs.

AM/FM Radios:

Function: AM/FM radios are a basic but essential communication tool, allowing you to receive news broadcasts, emergency alerts, and weather updates.

Application: Keep a battery-powered or hand-crank AM/FM radio in your shelter to stay informed about the external situation. Many radios also feature built-in emergency alert systems (EAS) that automatically broadcast critical information during a disaster.

Limitations: AM/FM radios are primarily one-way communication devices, meaning you can receive information but cannot send messages. Their effectiveness also depends on the availability of broadcasts and the condition of local radio infrastructure.

NOAA Weather Radios:

Function: NOAA weather radios provide continuous broadcasts of weather information and emergency alerts from the National Oceanic and Atmospheric Administration (NOAA).

Application: These radios are especially useful for receiving real-time updates on weather conditions, which can impact your safety decisions, such as when to leave the shelter.

Limitations: Similar to AM/FM radios, NOAA weather radios are one-way communication devices. Additionally, their utility depends on the continued operation of NOAA stations during a disaster.

Two-Way Radios (Walkie-Talkies):

Function: Two-way radios, commonly known as walkie-talkies, allow for real-time communication between two or more parties within a certain range.

Application: Use two-way radios for communication between shelter occupants or with nearby family members, neighbors, or other shelters. They are particularly useful for coordinating activities and relaying information when outside communication is limited.

Limitations: The range of two-way radios is limited, typically up to several miles depending on terrain and obstructions. Additionally, they require batteries, so it's important to have spares or rechargeable options.

Ham Radios (Amateur Radio):

Function: Ham radios are powerful communication devices that operate on various frequencies, allowing for long-distance communication even when traditional networks are down.

Application: Ham radios are ideal for reaching out to other amateur radio operators, emergency services, or coordinating with distant contacts. They can be used to gather information from a wide area, which is crucial for understanding the broader impact of a nuclear event.

Limitations: Ham radios require a license to operate, and using them effectively requires some technical knowledge and practice. Additionally, they can be expensive and bulky, requiring space and power sources in your shelter.

Satellite Phones:

Function: Satellite phones connect directly to satellites orbiting the Earth, providing communication capabilities even in areas where cellular networks are down or nonexistent.

Application: A satellite phone can be a lifeline for contacting emergency services or communicating with loved ones in remote areas. It is particularly useful in situations where all terrestrial communication networks have been disrupted.

Limitations: Satellite phones are expensive, and their usage can be costly due to high per-minute charges. They also require a clear view of the sky to maintain a connection, which can be challenging in certain shelter setups.

CB Radios (Citizen's Band Radio):

Function: CB radios operate on shortwave frequencies and allow for communication over short to medium distances without the need for a license.

Application: CB radios can be used to communicate with other nearby shelters, vehicles, or individuals. They are widely available and relatively easy to use.

Limitations: The range of CB radios is limited, typically around 3-20 miles depending on conditions. They are also prone to interference and are generally less reliable than ham radios for long-distance communication.

Mesh Networks:

Function: Mesh networks involve a series of interconnected devices that communicate with each other to form a network. These networks can operate independently of traditional internet and cellular networks.

Application: In a shelter scenario, mesh networks can be used to maintain communication between multiple shelters or households within a community, even if the broader communication infrastructure is down.

Limitations: Mesh networks require multiple devices to create a reliable network. Their range is typically limited, and they may require technical knowledge to set up and maintain.

Powering Communication Devices

In a fallout shelter, reliable power sources for your communication devices are essential. Without power, even the best communication systems are rendered useless.

Battery Power:

Primary Batteries: Many communication devices use primary (non-rechargeable) batteries. Stock up on extra batteries, and rotate your stock regularly to ensure freshness.

Rechargeable Batteries: Consider using rechargeable batteries for devices that support them. Pair these with a hand-crank generator, solar charger, or battery pack to maintain power.

Solar Power:

Solar Chargers: Solar chargers can be used to power communication devices, especially during extended shelter stays. Ensure your shelter has access to sunlight or consider setting up solar panels outside the shelter that can connect to your devices.

Solar-Powered Radios: Some radios come with built-in solar panels, providing a self-sustaining power source that can be a valuable asset in a long-term emergency.

Hand-Crank Generators:

Hand-Crank Radios: Many emergency radios are equipped with hand-crank generators, allowing you to generate power manually. This feature is invaluable when other power sources are unavailable.

Hand-Crank Chargers: Hand-crank chargers can be used to power small devices like two-way radios or cell phones in a pinch, though they require significant manual effort.

Backup Generators:

Portable Generators: A portable generator can provide power for multiple communication devices, as well as other essential systems in your shelter. Ensure that your generator is properly ventilated and that you have sufficient fuel stored.

Battery Packs: High-capacity battery packs can provide extended power for communication devices. Consider pre-charging these packs and keeping them ready for use.

Setting up a Communication Plan

A well-thought-out communication plan ensures that all shelter occupants understand how to use the available communication systems and know what to do in various scenarios.

Designating Roles:

Communication Officer: Assign a designated person in the shelter to manage and monitor communication systems. This person should be responsible for maintaining equipment, conducting regular checks, and coordinating communication efforts.

Training and Drills: Ensure that all shelter occupants are trained on how to use the communication devices, including changing frequencies, operating power sources, and responding to emergencies. Conduct regular drills to practice communication procedures.

Communication Protocols:

Check-In Schedule: Establish a regular schedule for checking in with external contacts or other shelters. This could be every hour, twice a day, or at intervals suited to your situation.

Code Words and Signals: Develop code words or signals to communicate specific messages quickly and discreetly. This can be useful in scenarios where you need to convey important information without revealing sensitive details.

Emergency Procedures: Clearly define what constitutes an emergency and how to respond. For example, if a PRD alarm indicates rising radiation levels, have a procedure in place for relaying this information and taking appropriate action.

Redundancy and Backup:

Multiple Devices: Ensure that you have more than one type of communication device available, in case one fails. For example, have both a ham radio and a satellite phone, as well as a backup power source for each.

Prearranged Contacts: Establish a list of prearranged contacts, including family members, local emergency services, and ham radio operators, who can be reached in an emergency. Ensure that all shelter occupants have a copy of this contact list.

Monitoring External Communication Channels

Staying informed about the external environment is crucial for making safe decisions about when to leave the shelter. Here are some strategies for effective monitoring:

Emergency Broadcasts:

EAS (Emergency Alert System): Regularly monitor the EAS on AM/FM radios for official updates on the situation. These broadcasts will provide critical information about radiation levels, evacuation orders, and other safety concerns.

NOAA Weather Alerts: Keep your NOAA weather radio on to receive alerts about weather conditions that could impact your safety, such as storms or changes in wind direction that might affect fallout patterns.

Ham Radio Networks:

Local and National Nets: Tune into local and national ham radio nets (networks) to gather information from other operators. Many ham radio enthusiasts participate in emergency response efforts and can provide valuable real-time data.

Emergency Frequencies: Familiarize yourself with the emergency frequencies used by ham radio operators and ensure your radio is programmed to access these frequencies. Regularly monitor these channels to stay informed about developments and to connect with other operators who may have critical information.

Shortwave Radio:

International Broadcasts: Shortwave radios can pick up international broadcasts, which might provide a broader perspective on the situation beyond your immediate area. This can be particularly valuable if local communication networks are down or unreliable.

Monitoring Global News: Tune into global news stations on shortwave radio to gather information about the extent of the nuclear event, international responses, and potential recovery efforts.

Citizen Reports:

CB Radio Channels: CB radio users often share local observations and updates. While the information on CB channels may be less reliable than official broadcasts, it can still provide insights into local conditions.

Online and Mesh Networks: If internet access is available through satellite or mesh networks, monitor social media, forums, and other online platforms for real-time updates from individuals in your area. Be cautious about the accuracy of this information, and cross-check with more reliable sources.

Maintaining Communication Systems over Time

Long-term use of communication systems in a fallout shelter requires ongoing maintenance to ensure reliability. Regular checks and proper storage are essential.

Regular Equipment Checks:

Power Sources: Test batteries, generators, and solar panels regularly to ensure they are fully charged and functioning properly. Replace or recharge batteries as needed to avoid power failures during critical times.

Signal Testing: Periodically test all communication devices to ensure they are transmitting and receiving signals correctly. This includes checking antenna connections, calibrating frequencies, and ensuring that microphones and speakers are working.

Filter and Protect: Keep dust, moisture, and radiation exposure to a minimum by storing communication devices in protective cases or containers when not in use. Consider using Faraday cages to protect sensitive electronics from electromagnetic pulses (EMPs).

Training and Skill Development:

Ongoing Training: Continuously train all shelter occupants on how to use the communication systems, especially if new devices are added. Practice drills should be a regular part of your shelter routine.

Ham Radio Skills: If you're using ham radios, invest time in learning the technical aspects of amateur radio operation. Consider obtaining a ham radio license and joining local ham radio clubs to develop your skills.

Documentation:

User Manuals: Keep all user manuals and operational guides for your communication devices in a safe, accessible location within the shelter. This will help troubleshoot issues and provide guidance if someone unfamiliar with the devices needs to operate them.

Logbooks: Maintain a logbook for recording communication activities, including times of contact, channels used, and key information received. This helps track communication patterns and ensures that critical information is documented.

Preparing for Post-Shelter Communication

When it's time to leave the shelter, having a communication plan in place is crucial for ensuring safety during the transition period.

Re-establishing Contact:

Emergency Contacts: Upon exiting the shelter, immediately reestablish contact with emergency services, family members, and other key contacts to inform them of your status and gather updated information about the external environment.

Coordination with Authorities: Coordinate with local authorities or emergency services to receive guidance on safe areas, ongoing hazards, and available resources. Use your most reliable communication device, such as a satellite phone or ham radio, for this purpose.

Monitoring Conditions:

Radiation Levels: Continue monitoring radiation levels as you leave the shelter. Use portable Geiger counters or PRDs to ensure that you're moving into areas with safe radiation levels.

Local and National Updates: Keep your AM/FM or shortwave radio on to receive updates about the situation in your area. Pay attention to any new advisories, warnings, or recovery efforts that might affect your next steps.

Long-Distance Communication:

Maintaining Contact: If you plan to travel to another location, use your communication devices to maintain contact with your destination. This could involve coordinating with family, friends, or emergency services along your route.

Navigating Disruptions: Be prepared for continued disruptions in communication networks as you move outside the shelter. Having multiple communication options and backup power sources will be essential for staying connected.

Effective communication is a cornerstone of survival in a fallout shelter. By equipping your shelter with a variety of communication systems, from basic AM/FM radios to advanced ham radios and satellite phones, you ensure that you can stay informed, coordinate with others, and call for help if needed. Proper training, regular maintenance, and a well-thought-out communication plan are essential to maximizing the effectiveness of these systems.

In a crisis, staying connected to the outside world and to those within your shelter can make the difference between life and death. With the right communication tools and strategies in place, you can navigate the challenges of a fallout scenario with confidence, ensuring that you and your loved ones remain safe and informed throughout the event and during the critical period of recovery afterward.

Power Sources: Keeping the Lights on Underground

Maintaining a reliable power supply in a fallout shelter is crucial for ensuring the functionality of critical systems, including ventilation, communication, lighting, and water purification. In the confined environment of a shelter, where you might need to stay for an extended period, keeping the lights on and the essential devices running is not just a matter of comfort but of survival. This chapter will explore the various power sources available for fallout shelters, discussing their advantages, limitations, and how to integrate them into your shelter's overall design.

The Importance of Power in a Fallout Shelter

Power is essential for the operation of many systems in a fallout shelter:

Lighting: Reliable lighting is necessary for daily activities, mental well-being, and safety. In a dark, enclosed space, proper lighting can help reduce stress and maintain morale.

Ventilation and Air Filtration: Power is needed to run ventilation systems that ensure a supply of clean air, remove carbon dioxide, and filter out radioactive particles and other contaminants.

Communication Systems: Maintaining contact with the outside world and receiving emergency broadcasts requires power for radios, satellite phones, and other communication devices.

Water and Sanitation: Power may be required for pumping water, operating purification systems, and managing sanitation facilities.

Heating and Cooling: In some climates, maintaining a safe temperature inside the shelter will require powered heating or cooling systems.

Primary Power Sources

Several primary power sources can be used in a fallout shelter, each with its own strengths and weaknesses. Selecting the right combination of power sources will depend on your specific needs, location, and budget.

Grid Power:

Function: The local electrical grid is the most common source of power for homes and buildings. In some cases, it may continue to operate for a time after a disaster.

Application: If grid power is still available, it can be used to run all essential systems in the shelter. However, reliance on grid power alone is risky, as it may fail during a nuclear event or subsequent emergencies.

Limitations: Grid power is highly vulnerable to disruption in a nuclear scenario, whether due to EMPs, infrastructure damage, or overloading. It should not be relied upon as the sole power source in a fallout shelter.

Generators:

Function: Generators convert fuel (gasoline, diesel, propane, or natural gas) into electricity, providing a reliable and portable power source.

Application: Generators are ideal for supplying power to critical systems during short-term outages. They can provide a substantial amount of power, making them suitable for running ventilation systems, lighting, and communication devices.

Limitations: Generators require a steady supply of fuel, which must be safely stored in or near the shelter. They also produce exhaust fumes that must be vented outside, requiring careful planning to avoid contaminating the shelter. Additionally, generators can be noisy, which might draw unwanted attention in a post-disaster environment.

Battery Backup Systems:

Function: Battery backup systems store electricity and provide power when the main source (such as the grid or a generator) is unavailable. They can be charged via grid power, generators, or renewable sources like solar panels.

Application: Batteries are crucial for maintaining power during brief outages and can serve as a bridge between longer-term power solutions. They are silent and produce no emissions, making them ideal for use inside the shelter.

Limitations: The capacity of battery systems is limited, depending on the size and type of batteries used. Over time, batteries will deplete and require recharging, so they should be used in conjunction with a method for replenishing their charge.

Renewable Power Sources

Renewable power sources are valuable for providing a continuous supply of energy during extended stays in a fallout shelter, especially when traditional fuel supplies might be limited or depleted.

Solar Power:

Function: Solar panels convert sunlight into electricity, which can be used directly or stored in batteries for later use.

Application: Solar power is an excellent renewable energy source for shelters, especially if the panels can be placed outside the shelter in a secure location. Solar power can be used to charge battery systems or directly power low-energy devices.

Limitations: The effectiveness of solar power depends on sunlight availability, which can be limited by weather conditions, fallout, or debris. Additionally, solar panels require significant space and may be vulnerable to damage or theft in a post-disaster scenario.

Wind Power:

Function: Wind turbines convert wind energy into electricity, providing a renewable source of power that can be used in combination with or as an alternative to solar power.

Application: Wind power can be a reliable energy source in areas with consistent wind patterns. Small wind turbines can be used to charge batteries or directly power shelter systems.

Limitations: Like solar power, wind power is dependent on environmental conditions. Wind turbines also require proper installation and maintenance, and their visibility may make them more susceptible to damage or theft.

Hydropower:

Function: Hydropower generates electricity from flowing or falling water, typically via a small turbine or waterwheel.

Application: Hydropower can be a reliable and continuous power source if your shelter is located near a flowing water source, such as a river or stream. It's particularly useful in remote or rural locations.

Limitations: Hydropower requires proximity to a suitable water source, which is not feasible for most urban or suburban shelters. Additionally, the equipment needed can be complex and may require regular maintenance.

Backup and Emergency Power Solutions

In addition to primary and renewable power sources, having backup and emergency power options is crucial for ensuring continuous operation during unexpected situations.

Hand-Crank Generators:

Function: Hand-crank generators convert manual effort into electricity, providing a small but immediate power source for essential devices.

Application: These generators are ideal for charging small electronics like radios, flashlights, or cell phones in an emergency. They are portable, easy to use, and do not rely on fuel or sunlight.

Limitations: The power output from hand-crank generators is limited, and they require continuous manual effort to produce electricity, making them unsuitable for powering larger systems or providing long-term energy.

Pedal-Powered Generators:

Function: Pedal-powered generators use the motion of pedaling to generate electricity, similar to how a bicycle dynamo works.

Application: Pedal generators can provide more sustained power than hand-crank devices and are useful for generating electricity for small devices or charging batteries. They also serve as a physical activity outlet, which can be beneficial for mental health during extended stays in the shelter.

Limitations: Like hand-crank generators, pedal generators require physical effort and are not suitable for powering large systems. They are best used as a supplementary power source.

Portable Power Banks:

Function: Portable power banks are rechargeable battery packs that store electricity for later use.

Application: Power banks are useful for charging small devices like phones, tablets, or radios. They are portable and can be pre-charged before entering the shelter.

Limitations: The capacity of power banks is limited, and once depleted, they need to be recharged from another power source. They are a good short-term solution but not suitable for long-term power needs.

Managing and Conserving Power

Efficient power management and conservation are critical in a fallout shelter, where resources may be limited, and the duration of your stay is uncertain.

Prioritizing Essential Systems:

Critical Devices First: Prioritize power for essential systems such as lighting, ventilation, communication devices, and water purification. Non-essential devices should be powered down to conserve energy.

Load Management: Use power strips with surge protectors to manage multiple devices from a single outlet. Turn off devices when not in use to reduce unnecessary power consumption.

Power-Saving Strategies:

Energy-Efficient Lighting: Use LED lights, which consume significantly less power than traditional incandescent bulbs, to illuminate the shelter. Consider using dimmable or task lighting to further reduce power usage.

Low-Power Modes: Set devices to low-power or energy-saving modes whenever possible. This reduces their power consumption without sacrificing functionality.

Scheduled Operation: Operate power-hungry devices, such as ventilation systems, on a scheduled basis rather than continuously. For example, run the ventilation system at regular intervals to refresh the air without depleting battery reserves.

Monitoring Power Usage:

Energy Meters: Install energy meters to monitor the power consumption of individual devices or the entire shelter. This helps identify where power is being used and allows for adjustments to conserve energy.

Battery Levels: Regularly check battery levels and plan recharging or fuel replenishment accordingly. Knowing your available power reserves helps in making informed decisions about energy usage.

Planning for Long-Term Power Needs

In situations where a stay in the shelter might extend for weeks or months, planning for long-term power needs is essential.

Expanding Renewable Power:

Solar and Wind Combinations: Consider combining solar and wind power to take advantage of different weather conditions. This diversification helps ensure a more consistent power supply.

Battery Storage: Invest in high-capacity battery storage systems that can store excess energy generated during peak production times. This stored energy can be used during periods of low solar or wind activity.

Fuel Management for Generators:

Fuel Storage: Safely store an adequate supply of fuel for your generator, keeping in mind that different fuels have different shelf lives. Rotate your fuel stock regularly to ensure it remains usable.

Fuel Efficiency: Use fuel-efficient generators and only run them when necessary. Consider using your generator during peak power needs, such as when charging batteries, running critical systems, or cooking. When not in use, rely on stored energy in batteries or renewable sources to conserve fuel.

Alternative Energy Sources:

Biofuel Generators: If you anticipate a long-term stay in the shelter and have access to organic waste, consider a biofuel generator. These systems convert organic materials like food scraps or plant waste into usable fuel for powering a generator.

Human-Powered Generators: While not suitable for high-power needs, human-powered generators, such as pedal generators, can provide a renewable source of energy for small devices or to supplement other power sources. This approach also doubles as a form of exercise, which can help maintain physical and mental well-being during an extended shelter stay.

Upgrading Power Systems:

Inverter Systems: Consider installing a power inverter that can convert DC power (from batteries or solar panels) to AC power, which is used by most household appliances. An inverter system allows you to run more devices and makes your power setup more flexible.

Hybrid Systems: A hybrid power system that combines multiple sources—solar, wind, generator, and batteries—can provide a more robust and reliable energy solution. These systems automatically switch between power sources depending on availability, ensuring continuous operation of critical systems.

Ensuring Safety and Reducing Risks

Safety is paramount when managing power in a fallout shelter, especially when using fuel-based generators or large battery systems.

Proper Ventilation for Generators:

Exhaust Management: Ensure that fuel-based generators are used in a well-ventilated area outside the main shelter space. Properly vent exhaust fumes to prevent the buildup of carbon monoxide, which can be deadly in confined spaces.

Carbon Monoxide Detectors: Install carbon monoxide detectors in the shelter, especially if you're using a generator. These detectors can provide early warning if dangerous fumes are detected, allowing you to take immediate action.

Battery Safety:

Safe Storage: Store batteries in a cool, dry place, away from direct sunlight or heat sources. Ensure that they are kept off the ground and in a location where they cannot be easily damaged.

Fire Prevention: Batteries, especially large storage systems, can pose a fire risk. Install fire extinguishers rated for electrical fires and avoid overloading circuits. Regularly inspect batteries for signs of wear, corrosion, or damage, and replace them as needed.

Generator Fuel Storage:

Fuel Storage Guidelines: Store fuel in approved containers, clearly labeled, and kept in a well-ventilated, secure area away from living spaces. Avoid storing too much fuel inside the shelter itself to reduce fire hazards.

Fuel Stabilizers: Use fuel stabilizers to extend the shelf life of stored gasoline or diesel. This is particularly important if you anticipate needing to store fuel for an extended period.

Final Preparations and Planning

Before considering your power system ready for use, it's essential to conduct a thorough review and final preparations to ensure everything is in place and functioning correctly.

System Testing:

Full Load Testing: Test your entire power system under full load conditions to ensure it can handle the demands of all critical systems running simultaneously. This includes running the generator, testing battery backup systems, and checking the operation of renewable energy sources.

Redundancy Check: Verify that all backup and emergency power systems, such as hand-crank generators and portable power banks, are fully functional and ready for use. Ensure you have spare parts and supplies for maintaining these systems.

Training and Drills:

Operational Training: Train all shelter occupants on the operation of the power systems, including how to start and stop the generator, switch between power sources, and monitor energy usage. Ensure everyone understands how to conserve power and what to do in case of a power failure.

Emergency Drills: Conduct drills to simulate power outages and practice switching to backup systems. These drills should also include scenarios such as recharging batteries with solar panels or manually operating ventilation systems.

Documentation:

Power System Manual: Create a manual that details the operation, maintenance, and troubleshooting of your shelter's power systems. Include contact information for service providers or manufacturers in case of technical issues.

Maintenance Log: Keep a log of all maintenance activities, including fuel usage, battery checks, and system tests. This log will help track the condition of your power systems and ensure that they remain in optimal working order.

A reliable and well-planned power system is essential for the survival and comfort of a fallout shelter's occupants. By understanding the different power sources available—whether grid power, generators, renewable energy, or emergency backups—you can design a system that meets your shelter's needs, both for short-term emergencies and long-term stays. Efficient power management, regular maintenance, and safety precautions are key to ensuring that your shelter remains a safe and liveable space throughout a crisis. By preparing for all potential scenarios and having multiple power options at your disposal, you can ensure that the lights stay on, the air remains clean, and communication lines remain open, providing security and peace of mind during even the most challenging circumstances.

Water Storage: Ensuring a Safe and Reliable Supply

Water is one of the most critical resources in a fallout shelter, essential for drinking, cooking, hygiene, and sanitation. In a confined environment where external water sources may be contaminated or inaccessible, having a safe and reliable supply of water is crucial for survival. This chapter will explore the strategies for water storage in a fallout shelter, including the types of containers, methods for purification, and techniques for managing and conserving water over an extended period.

The Importance of Water in a Fallout Shelter

Water is vital for sustaining life and maintaining health in a fallout shelter:

Drinking Water: The human body needs a regular intake of clean water to function properly. Dehydration can lead to serious health problems, including kidney failure, heatstroke, and impaired cognitive function.

Cooking and Food Preparation: Water is required for preparing food, especially if you're relying on stored or dehydrated foods that need to be rehydrated or cooked.

Hygiene and Sanitation: Personal hygiene, such as washing hands and cleaning surfaces, requires water to prevent the spread of germs and maintain a healthy environment. Additionally, water is needed for operating sanitation systems, like flushing toilets or managing waste.

Emergency Medical Care: In the event of an injury or illness, water is needed for cleaning wounds, administering first aid, and possibly for taking medication.

Estimating Water Needs

Before you begin storing water, it's important to estimate how much water you'll need for your shelter's occupants over the expected duration of your stay.

Basic Water Needs:

Drinking: The general recommendation is to store at least one gallon of water per person per day. This includes water for drinking, food preparation, and minimal hygiene.

Hygiene and Sanitation: For basic hygiene and sanitation needs, add an additional half to one gallon of water per person per day. This will cover essential activities like hand washing, surface cleaning, and waste management.

Emergency Reserves: It's advisable to store extra water beyond the minimum estimates to account for unexpected situations, such as a longer-than-expected stay, spills, or additional occupants.

Duration of Shelter Stay:

Short-Term: For a shelter stay of up to two weeks, calculate the total water requirement by multiplying the daily needs by the number of days. For example, a family of four staying for 14 days would require at least 56 gallons of water for drinking and hygiene.

Long-Term: For extended stays of several months, consider not only the total amount of water required but also how to manage, conserve, and potentially replenish your water supply.

Water Storage Containers

Choosing the right containers for water storage is essential for maintaining water quality and ensuring ease of access in the shelter.

Types of Water Containers:

Food-Grade Plastic Containers: These are among the most common and practical options for water storage. Food-grade plastic containers are durable, lightweight, and come in various sizes, from small jugs to large barrels. Ensure that the containers are made from BPA-free plastic to avoid chemical contamination.

Water Storage Tanks: Large-capacity tanks, typically made from polyethylene or other durable materials, can store significant amounts of water. These tanks can be placed inside or outside the shelter, depending on space availability and design considerations.

Collapsible Water Containers: These containers are made from flexible materials and can be folded when not in use, making them ideal for conserving space in a shelter. They are useful as supplementary storage or for carrying water from an external source.

Water Bricks: Water bricks are stackable, interlocking containers that are easy to store and organize in a shelter. They typically hold 3 to 5 gallons of water each and are designed to maximize space efficiency.

Considerations for Container Selection:

Durability: Choose containers that are strong enough to withstand long-term storage and any potential impact or pressure in the shelter.

Portability: While larger containers are more efficient for storing large quantities of water, smaller, portable containers are easier to move and manage, especially if you need to transport water from one area to another.

Sealing: Ensure that all containers have tight-fitting lids or caps to prevent contamination and evaporation. Consider using containers with spigots or taps for easy dispensing without the need to open the entire container.

Space Utilization: Plan the layout of your shelter to maximize the use of space for water storage. Stackable or modular containers can help you store more water in a confined area.

Water Purification and Treatment

Even if you store clean water, it's important to have the means to purify and treat water to ensure it remains safe to drink over time.

Purification Methods:

Boiling: Boiling water for at least one minute (or three minutes at higher altitudes) is an effective method to kill bacteria, viruses, and parasites. This method requires a heat source and should be used when possible.

Chemical Treatment: Chemical disinfectants like chlorine bleach (sodium hypochlorite) or iodine tablets can be used to treat water. Add 8 drops of bleach per gallon of water and let it sit for at least 30 minutes before use. For iodine, follow the instructions provided with the tablets.

Filtration: Water filters, such as portable pump filters, gravity filters, or straw filters, can remove bacteria, protozoa, and some viruses from water. Choose a filter with a pore size of 0.2 microns or smaller for best results.

UV Purification: UV water purifiers use ultraviolet light to kill microorganisms in water. These devices are compact and easy to use but require batteries or another power source.

Activated Carbon Filters: These filters can remove chemicals, odors, and some heavy metals from water. They are often used in combination with other purification methods to improve taste and reduce chemical contaminants.

Storage Stability:

Water Preservers: Consider using water preservatives, such as commercially available water stabilizers, to extend the shelf life of stored water. These products are typically added to water to prevent the growth of bacteria, algae, and other microorganisms for up to five years.

Regular Rotation: Even with preservatives, it's a good practice to rotate your water supply every six to twelve months. Use the old water for non-potable purposes, such as cleaning or gardening, and replace it with fresh water.

Water Testing:

Test Kits: Use water testing kits to periodically check the quality of your stored water. These kits can test for bacteria, chlorine levels, pH, and other contaminants, helping you ensure the safety of your water supply.

Visual Inspection: Regularly inspect water containers for any signs of contamination, such as cloudiness, discoloration, or unusual odors. If the water appears compromised, treat or replace it immediately.

Managing and Conserving Water

In a fallout shelter, where water may be a limited resource, managing and conserving it is essential for long-term survival.

Water Rationing:

Daily Allotment: Establish a daily water ration for each shelter occupant based on the estimated needs. Stick to this ration to ensure that the water supply lasts for the expected duration of the shelter stay.

Non-Potable Water: Use non-potable water for tasks that don't require drinking-quality water, such as flushing toilets, washing clothes, or cleaning floors. This can include water collected from rainfall, greywater, or even melted ice.

Recycling and Reusing Water:

Greywater Systems: Consider setting up a simple greywater recycling system, where water from sinks, showers, or handwashing is collected and reused for non-potable purposes, such as flushing toilets or irrigating plants in a survival garden. Ensure that greywater is not contaminated with harmful chemicals or pathogens before reuse.

Rainwater Harvesting:

Collection Systems: If your shelter is above ground or has access to an exterior area, consider setting up a rainwater harvesting system. This involves collecting and storing rainwater from roofs or other surfaces. Use gutters and downspouts to direct water into storage tanks or barrels.

Purification: Rainwater should be purified before drinking, as it can collect contaminants from the air or the collection surface. Filter the water and treat it with chemical disinfectants or boil it before use.

Water Discipline:

Conservation Practices: Encourage all shelter occupants to practice water conservation by taking shorter washes, using minimal water for cooking, and avoiding unnecessary water use.

Monitoring Usage: Keep track of daily water usage and adjust rations as necessary to ensure that your supply lasts. Regularly review your water management practices to identify any areas where further conservation can be achieved.

Emergency Water Sources

In the event that your stored water supply runs low or becomes contaminated, it's important to have plans in place for accessing additional water.

Alternative Water Sources:

Natural Water Bodies: Identify any nearby lakes, rivers, or streams that could serve as emergency water sources. Be aware that these sources are likely to be contaminated in the aftermath of a nuclear event, so rigorous purification is necessary.

Underground Wells: If your shelter has access to a well, ensure that it is properly sealed and protected from surface contamination. You may need a manual pump if the well is not connected to an electric pump.

Melting Snow or Ice: In colder climates, snow and ice can be melted for drinking water. Always purify the water before drinking, as it can still contain contaminants.

Portable Water Purifiers:

Portable Filters: Keep portable water filters on hand for purifying water from external sources. These filters should be capable of removing bacteria, protozoa, and viruses.

Chemical Treatment Kits: Stock chemical treatment kits that include chlorine or iodine tablets, which can be used to treat water from natural sources.

Water Extraction Techniques:

Solar Still: A solar still is a device that uses the heat of the sun to evaporate water, leaving contaminants behind, and then condenses the water vapor back into liquid. It's a useful method for extracting drinkable water from contaminated sources or even from moist soil.

Dew Collection: In environments with significant humidity, collect dew that forms on surfaces overnight. Spread out a clean plastic sheet or use a tarp to capture dew, then funnel it into a container for purification.

Final Preparations and Maintenance

Before your shelter is considered ready, ensure that your water storage systems are fully prepared and that maintenance plans are in place.

Pre-Shelter Checklist:

Container Inspection: Inspect all water containers for leaks, cracks, or signs of wear. Replace any compromised containers to prevent water loss.

Fill and Seal: Fill all containers with clean, potable water and seal them tightly. Consider adding water preservers if you anticipate long-term storage.

Accessibility: Arrange water containers in a manner that allows easy access and distribution. Ensure that the containers you will use first are the most accessible.

Ongoing Maintenance:

Regular Checks: Periodically check water containers for signs of contamination, such as discoloration, sediment, or odors. Test water quality using test kits and replace or treat water as needed.

Rotation: Establish a water rotation schedule to ensure that stored water remains fresh. Rotate your water supply every six to twelve months, using older water for non-potable purposes and replacing it with fresh water.

Training and Education:

Water Management Training: Educate all shelter occupants on the importance of water conservation and the proper use of water storage and purification systems. Regularly review and practice emergency water sourcing and purification techniques.

Emergency Drills: Conduct drills that simulate a water shortage or contamination event, allowing occupants to practice emergency water collection and purification procedures.

Water is an indispensable resource in a fallout shelter, and proper storage, management, and purification are essential for survival. By carefully estimating your water needs, selecting appropriate storage containers, and implementing effective purification methods, you can ensure a safe and reliable water supply throughout your stay in the shelter. In addition to storing water, planning for alternative water sources and practicing conservation techniques will help you manage this precious resource during an extended crisis. With thorough preparation and regular maintenance, you can provide yourself and your shelter occupants with the water needed to stay healthy, hydrated, and secure during the most challenging of circumstances.

Food Storage: Long-Term Survival Planning

Food storage is a critical component of long-term survival in a fallout shelter. Properly stored and managed food supplies can sustain shelter occupants for weeks or even months, ensuring they have the necessary nutrients and energy to endure the hardships of a nuclear fallout scenario. This chapter will explore the strategies for long-term food storage, focusing on selecting the right types of food, storing them effectively, and managing the supply to prevent spoilage and waste.

The Importance of Food Storage in a Fallout Shelter

Food is essential for maintaining physical health, mental well-being, and energy levels during an extended stay in a fallout shelter:

Nutritional Needs: A balanced diet that includes carbohydrates, proteins, fats, vitamins, and minerals is crucial for keeping your body functioning properly under stressful conditions.

Morale and Comfort: Access to familiar and comforting foods can help boost morale and provide a sense of normalcy in a crisis situation.

Sustainability: Adequate food storage allows for long-term sustainability, reducing the need to leave the shelter in search of supplies and minimizing the risk of exposure to dangerous external conditions.

Estimating Food Needs

Before you begin storing food, it's important to estimate how much you'll need based on the number of occupants and the expected duration of your stay in the shelter.

Daily Caloric Intake:

Calorie Requirements: The average adult requires about 2,000 to 2,500 calories per day to maintain body weight and energy levels. However, caloric needs may vary based on age, gender, activity level, and overall health.

Adjustments for Shelter Life: In a shelter environment, where physical activity may be limited, caloric needs might decrease slightly. However, stress and the need for warmth may increase caloric requirements. Plan for at least 2,000 calories per person per day.

Duration of Shelter Stay:

Short-Term: For stays up to two weeks, multiply the daily caloric requirement by the number of days and the number of occupants. For example, a family of four staying for 14 days would need at least 112,000 calories in total.

Long-Term: For extended stays of several months, plan for a variety of food types to prevent diet fatigue and ensure nutritional balance. Consider the shelf life of different foods and plan to rotate supplies as necessary.

Types of Long-Term Food Storage

Choosing the right types of food for long-term storage is crucial for ensuring that your supplies remain safe, edible, and nutritious over time.

Non-Perishable Foods:

Canned Goods: Canned foods, such as vegetables, fruits, meats, and soups, are staples of long-term food storage. They have a long shelf life, typically ranging from 2 to 5 years, and are easy to store. Look for low-sodium and nutrient-rich options.

Dried Foods: Dried foods, such as beans, rice, pasta, and lentils, are lightweight, have a long shelf life, and are calorie-dense. They require water for preparation, so ensure you have enough water stored to rehydrate these foods.

Freeze-Dried Foods: Freeze-dried meals and ingredients are ideal for long-term storage, with a shelf life of up to 25 years. They are lightweight, easy to prepare (requiring only hot water), and retain much of their nutritional value.

Dehydrated Foods: Dehydrated fruits, vegetables, and meats have had most of their water content removed, making them lightweight and shelf-stable. They can be rehydrated or eaten as-is, depending on the food type.

Grains and Legumes:

Whole Grains: Whole grains, such as wheat berries, oats, and barley, can be stored for several years if kept in a cool, dry place. They are nutrient-rich and versatile, but require preparation and cooking.

Legumes: Beans, lentils, and peas are excellent sources of protein and fiber. They have a long shelf life and can be stored in bulk. However, they require soaking and cooking, so plan for adequate water and fuel.

Ready-to-Eat Foods:

MREs (Meals Ready-to-Eat): MREs are self-contained, pre-cooked meals that are designed for military use and emergency situations. They have a shelf life of 3 to 5 years and do not require additional water or cooking, making them ideal for situations where resources are limited.

Energy Bars and Snacks: High-calorie energy bars, nuts, and trail mix are convenient, ready-to-eat options that provide quick energy. They have a relatively long shelf life and are easy to store.

Comfort Foods:

Chocolate and Sweets: Including some comfort foods, like chocolate, hard candy, or instant coffee, can help boost morale during an extended stay in the shelter. These items generally have a long shelf life and can be stored easily.

Spices and Condiments: Basic spices, salt, sugar, and condiments can enhance the flavor of stored foods and make meals more enjoyable. Store them in airtight containers to prevent moisture and contamination.

Food Storage Containers and Methods

Proper storage techniques are essential for preserving the quality and safety of your food supply over the long term.

Storage Containers:

Mylar Bags: Mylar bags are an excellent option for storing dried foods, grains, and legumes. They are durable, light-resistant, and can be sealed with an oxygen absorber to extend shelf life.

Vacuum-Sealed Bags: Vacuum-sealing removes air from the bag, reducing the risk of spoilage and extending the shelf life of foods like dried meats, nuts, and grains. Use vacuum-sealed bags for foods that are sensitive to moisture and oxygen.

Plastic Buckets: Food-grade plastic buckets with airtight lids are ideal for storing bulk items like grains, rice, and beans. They protect against moisture, pests, and physical damage. Consider using Mylar bags inside the buckets for added protection.

Glass Jars: Glass jars with airtight lids are suitable for storing smaller quantities of dried foods, spices, and preserved goods. They are impermeable to air and moisture but should be stored in a dark place to protect contents from light.

Storage Environment:

Cool and Dry: Store food in a cool, dry place, ideally at a temperature between 50°F and 70°F. High temperatures can cause foods to spoil more quickly, while moisture can lead to mold, bacteria growth, and contamination.

Darkness: Protect food from light exposure, which can degrade certain nutrients and spoil food over time. Store food in opaque containers or in a dark room, such as a basement or cellar.

Pest Control: Ensure that food storage areas are pest-proof by using sealed containers and keeping the area clean and free of crumbs or spilled food. Consider using diatomaceous earth or bay leaves to deter insects in storage areas.

Rotating and Managing Food Supplies

Managing your food supply effectively involves regular rotation and inventory control to prevent waste and ensure that your food remains safe and nutritious.

First-In, First-Out (FIFO):

Rotation Method: Practice the FIFO method, where the oldest food items are used first, and new items are added to the back of your storage. This ensures that nothing goes to waste and that your supply is always as fresh as possible.

Labeling: Clearly label all food containers with the date of purchase or storage, as well as the expiration date if available. This makes it easier to keep track of what needs to be used first.

Inventory Management:

Regular Inventory Checks: Conduct regular checks of your food inventory, noting any items that are nearing their expiration date or showing signs of spoilage. Adjust your food plan accordingly to use up older items first.

Restocking Plan: Develop a plan for restocking your food supplies, taking into account the shelf life of different items and your anticipated needs. If possible, replace items as they are used to maintain a steady supply.

Dealing with Spoilage:

Identifying Spoiled Food: Learn to identify signs of spoilage, such as unusual odors, mold, discoloration, or off-taste. Dispose of any food that shows signs of spoilage to prevent contamination of other stored foods.

Emergency Substitutions: Have a plan for substituting essential food items if your primary supply becomes spoiled or depleted. For example, if grains spoil, consider using stored legumes or pasta as a substitute carbohydrate source.

Supplementing Your Food Supply

In a long-term shelter situation, supplementing your stored food supply can help extend your resources and provide fresh nutrients.

Survival Gardening:

Indoor Gardens: If space and resources allow, consider setting up an indoor survival garden. Grow fast-growing, nutrient-rich plants like leafy greens, herbs, and sprouts, which can provide fresh vitamins and minerals.

Container Gardening: Use containers or grow bags to cultivate small amounts of vegetables or herbs inside the shelter. Even a small garden can make a significant difference in your diet over time.

Foraging and Hunting:

Local Foraging: If it is safe to leave the shelter, consider foraging for edible plants, berries, and nuts in the surrounding area. Learn to identify safe, nutrient-rich wild foods that can supplement your diet.

Small-Game Hunting: In a long-term scenario, hunting small game like rabbits or birds could provide additional protein. Ensure that you have the necessary tools and skills to safely hunt and prepare wild game.

Bartering:

Trade with Others: If you are part of a larger community of shelters or survivors, consider bartering for additional food supplies. Trading surplus items like canned goods, preserved foods, or comfort items can help diversify your diet and acquire foods that you might not have stored in sufficient quantities.

Food Preservation Techniques:

Canning and Pickling: If you have access to fresh produce or meats, consider canning or pickling to preserve these items for long-term storage. This requires proper equipment and knowledge, but it can significantly extend the shelf life of perishable foods.

Drying and Smoking: Drying fruits, vegetables, or meats and smoking meats are effective ways to preserve food without refrigeration. These methods reduce moisture content, preventing bacterial growth and extending shelf life.

Preparing Meals in a Fallout Shelter

Meal preparation in a fallout shelter may differ from your normal routine due to limited resources and the need for conservation.

Simplified Cooking:

One-Pot Meals: Focus on preparing one-pot meals that combine multiple ingredients into a single dish. This method conserves water and fuel, reduces clean-up, and ensures that you're getting a balanced meal.

Minimal Cooking: Choose recipes that require minimal cooking time or can be prepared with boiling water. Quick-cooking grains, freeze-dried meals, and MREs are ideal for this purpose.

Alternative Cooking Methods:

Portable Stoves: Use portable stoves or camping stoves that can be fueled by propane, butane, or alcohol. Ensure that your shelter is well-ventilated when using these stoves to prevent carbon monoxide buildup.

Solar Ovens: If your shelter has access to sunlight, consider using a solar oven to cook or reheat meals. Solar ovens are an energy-efficient and safe cooking option, though they require direct sunlight to function effectively.

Fuel Conservation: Conserve fuel by batch cooking meals, preparing food in larger quantities and storing leftovers for later use. This approach reduces the need for frequent cooking and saves valuable resources.

Creative Food Use:

Reusing Ingredients: Make the most of every ingredient by reusing leftovers and scraps. For example, vegetable peelings can be used to make broth, and stale bread can be turned into croutons or breadcrumbs.

Stretching Supplies: Extend your food supplies by adding fillers like rice, beans, or pasta to meals. These ingredients are calorie-dense and can help make other, more limited ingredients go further.

Final Preparations and Maintenance

Before considering your food storage complete, ensure that your supplies are fully organized, secure, and ready for long-term use.

Organizing Supplies:

Shelving and Storage: Arrange your food supplies on sturdy shelves, with the heaviest items on the lower shelves to prevent accidents. Ensure that all items are easily accessible and that nothing is hidden or difficult to reach.

Category Grouping: Group similar items together (e.g., canned goods, grains, snacks) to make it easier to locate what you need. Label shelves or containers for quick identification.

Securing Food Storage:

Pest Control: Take steps to secure your food storage against pests by using airtight containers, regularly cleaning storage areas, and checking for signs of infestation. Keep traps or pest deterrents on hand as an additional precaution.

Temperature Control: Ensure that your food storage area maintains a stable, cool temperature, ideally between 50°F and 70°F. If necessary, insulate the storage area or use a thermometer to monitor conditions.

Regular Inventory and Rotation:

Routine Checks: Perform routine checks of your food inventory, looking for any signs of spoilage, damage, or expiration. Keep track of what's being used and what needs to be replaced.

Rotating Stock: Implement a regular rotation schedule to use up older items first and keep your food supply fresh. Consider marking the calendar with reminder dates for rotation and inventory checks.

Training and Education:

Meal Planning: Educate shelter occupants on how to plan meals using the stored food supplies. Teach them how to prepare basic meals, conserve resources, and recognize signs of food spoilage.

Emergency Drills: Conduct drills that simulate scenarios like food shortages or the need to ration supplies. These drills will help ensure that everyone knows how to manage and conserve food during an extended shelter stay.

Proper food storage is essential for long-term survival in a fallout shelter. By carefully selecting non-perishable and nutrient-dense foods, using appropriate storage containers, and maintaining an organized and secure food supply, you can ensure that your shelter's occupants have the sustenance they need during an extended crisis.

Managing your food supply effectively through regular rotation, inventory control, and creative meal preparation will help prevent waste and ensure that your resources last as long as necessary. Supplementing stored food with foraged or grown foods, along with the ability to preserve fresh items, can further extend your supplies and provide much-needed variety. With thorough preparation and ongoing maintenance, your food storage plan will provide a reliable source of nourishment, helping you and your loved ones stay healthy, strong, and resilient in the face of a nuclear fallout scenario.

Waste Management in a Sealed Environment

Effective waste management in a sealed environment, such as a fallout shelter, is critical to maintaining health, hygiene, and comfort during an extended stay. Without proper waste disposal systems, the shelter could quickly become unsanitary, leading to the spread of disease and compromising the safety and well-being of its occupants. This chapter will explore the strategies and systems for managing human waste, garbage, and other refuse in a sealed environment, ensuring that your shelter remains liveable and hygienic over the long term.

The Importance of Waste Management in a Fallout Shelter

Proper waste management is essential for several reasons:

Hygiene and Health: Accumulated waste can lead to the growth of harmful bacteria, viruses, and parasites, increasing the risk of infections and illness. Effective waste management helps prevent these health hazards.

Odor Control: Waste generates unpleasant odors that can make the shelter environment unbearable. Managing waste efficiently helps to control and minimize odors, contributing to a more comfortable living space.

Pest Control: Unmanaged waste can attract pests such as insects and rodents, which can further spread disease and contaminate food supplies. Keeping the shelter clean and waste-free reduces the likelihood of pest infestations.

Mental Well-being: A clean and organized shelter environment supports the mental well-being of its occupants, reducing stress and helping maintain a sense of normalcy during a crisis.

Managing Human Waste

One of the most important aspects of waste management in a sealed environment is the safe disposal of human waste. Several options are available, depending on your shelter's setup and resources.

Portable Toilets:

Chemical Toilets: Portable chemical toilets use chemicals to break down waste and control odors. These toilets are compact, easy to use, and can be sealed tightly to prevent leaks and smells. However, they require regular emptying and chemical refills.

Composting Toilets: Composting toilets decompose human waste into compost through a natural process involving heat, moisture, and microorganisms. These toilets are environmentally friendly and reduce the need for frequent waste disposal, but they require proper ventilation and maintenance.

Bucket Toilets: A simple and inexpensive option is the bucket toilet, which uses a bucket lined with heavy-duty plastic bags. After use, the bag is sealed and stored until it can be safely disposed of. This method is straightforward but requires careful handling and regular disposal of waste bags.

Ventilation and Odor Control:

Ventilation Systems: Ensure that your waste disposal area is well-ventilated to prevent the buildup of odors and harmful gases like ammonia and methane. If using a composting toilet, proper ventilation is essential for the composting process.

Absorbent Materials: Use absorbent materials like sawdust, peat moss, or cat litter to cover waste after each use. This helps absorb moisture, control odors, and aids in the composting process if applicable.

Odor Neutralizers: Keep odor-neutralizing products like activated charcoal, baking soda, or commercial odor absorbers in the waste area to help manage smells.

Waste Disposal:

Sealing and Storing Waste: If using a bucket or chemical toilet, waste should be sealed in heavy-duty plastic bags or containers after each use. Store sealed waste in a designated area away from living spaces until it can be safely removed or disposed of.

Periodic Disposal: Plan for the periodic disposal of accumulated waste, especially if your stay in the shelter extends beyond a few weeks. This may involve burying waste away from the shelter if it is safe to exit, or using a secondary, more permanent waste disposal solution.

Managing Garbage and Non-Human Waste

In addition to human waste, you'll need to manage other types of waste, including food scraps, packaging, and general garbage.

Segregation of Waste:

Organic vs. Inorganic Waste: Separate organic waste (food scraps, paper) from inorganic waste (plastics, metals) to manage disposal more effectively. Organic waste can be composted or stored separately to reduce odors and pests.

Hazardous Waste: Identify and segregate any hazardous waste, such as batteries, chemicals, or medical supplies, which require special handling and disposal. Store hazardous waste in clearly labeled containers away from food and water supplies.

Garbage Collection and Storage:

Trash Bags and Bins: Use heavy-duty trash bags and bins with tight-fitting lids to collect and store garbage. Regularly remove and seal trash bags to prevent odors and contamination.

Compression and Minimization: Compress waste to reduce its volume and save space. For example, flatten boxes, crush cans, and compress plastic bottles before storing them in trash bags.

Waste Reduction: Minimize waste production by reusing containers, avoiding unnecessary packaging, and composting organic material when possible.

Composting Organic Waste:

Indoor Composting: If space and resources allow, set up a small indoor composting system using a sealed container or bin. Add organic waste like food scraps, paper, and compostable materials, along with composting agents like worms or compost starters. Turn the compost regularly to aid decomposition and reduce odors.

Compost Management: Keep the composting area well-ventilated and monitor moisture levels to prevent it from becoming too wet or dry. Properly managed compost can be used to fertilize an indoor garden or stored for future use.

Water Waste Management

Managing greywater (wastewater from sinks, showers, or washing) and blackwater (wastewater from toilets) is crucial in a sealed environment to prevent contamination and maintain hygiene.

Greywater Reuse:

Collection and Filtration: Collect greywater in containers or buckets for reuse in non-potable applications, such as flushing toilets or watering plants. Filter the water to remove large particles and impurities before reuse.

Greywater Storage: Store greywater in sealed containers if immediate reuse is not possible. Ensure that storage containers are labeled and kept away from potable water supplies to avoid cross-contamination.

Blackwater Disposal:

Separation from Greywater: Always keep blackwater (contaminated with human waste) separate from greywater. Blackwater poses a higher risk of contamination and should be handled with greater care.

Safe Storage: Store blackwater in sealed, durable containers designed for hazardous waste. Ensure that the storage area is well-ventilated and regularly monitored for leaks or damage.

Emergency Disposal: If it becomes necessary to dispose of blackwater during an extended shelter stay, do so safely and far from the shelter to prevent contamination. Burying blackwater in a deep pit away from water sources and living areas is one method, if safe to exit the shelter.

Hygiene and Sanitation Practices

Maintaining personal hygiene and keeping the shelter clean are vital to preventing the spread of disease in a sealed environment.

Personal Hygiene:

Handwashing: Regular handwashing with soap and water is essential, especially after using the toilet, handling waste, or before eating. If water is limited, use alcohol-based hand sanitizers.

Body Cleansing: Use wet wipes, sponge baths, or water-efficient washing methods to maintain personal cleanliness. Ensure that greywater from washing is collected and properly managed.

Dental Hygiene: Practice good dental hygiene by brushing teeth regularly with a minimal amount of water. Consider storing waterless toothbrushes or oral hygiene wipes as a backup.

Shelter Cleaning:

Surface Disinfection: Regularly clean and disinfect surfaces, especially those in contact with food or waste. Use disinfectants like bleach solutions or commercial cleaners to kill bacteria and viruses.

Waste Area Maintenance: Clean and disinfect the waste disposal area frequently to prevent odors and contamination. Ensure that waste containers are regularly emptied, cleaned, and sealed.

Laundry Management:

Limited Water Usage: If water is limited, prioritize washing essential clothing and linens. Use minimal water and biodegradable soap, and consider reusing greywater for washing purposes.

Drying Clothes: Air-dry clothes on a line or drying rack inside the shelter to conserve energy and avoid excess moisture buildup. Ensure that the drying area is well-ventilated to prevent mold growth.

Final Preparations and Monitoring

Before finalizing your waste management plans, ensure that all systems and procedures are in place and that shelter occupants are trained and prepared.

System Setup:

Waste Disposal Stations: Set up designated waste disposal stations for human waste, garbage, and greywater in different areas of the shelter. Clearly label each station and provide instructions for proper use.

Emergency Kits: Prepare emergency waste management kits that include extra trash bags, disinfectants, gloves, and other essential supplies. Store these kits in easily accessible locations.

Training and Drills:

Waste Management Training: Train all shelter occupants on proper waste management procedures, including the use of portable toilets, segregation of waste, and emergency disposal methods.

Hygiene Drills: Conduct drills that simulate waste management scenarios, such as a full waste container or a malfunctioning toilet. These drills help ensure that everyone knows how to respond and maintain hygiene in an emergency.

Monitoring and Maintenance:

Regular Inspections: Conduct regular inspections of waste management systems to check for leaks, odors, or signs of contamination. Address any issues immediately to prevent them from worsening.

Supply Check: Monitor the availability of waste management supplies, such as trash bags, disinfectants, and chemicals for toilets. Restock as necessary to ensure you are always prepared.

Effective waste management in a sealed environment is essential for maintaining a safe, sanitary, and liveable space during an extended stay in a fallout shelter. By setting up proper systems for the disposal and management of human waste, garbage, and water waste, you can prevent the spread of disease, control odors, and keep pests at bay.

Thorough preparation, regular maintenance, and adherence to hygiene practices will help ensure that your shelter remains a healthy environment for all occupants. With a well-thought-out waste management plan in place, you can focus on survival and well-being, knowing that this critical aspect of shelter life is under control.

Psychological Preparation for Long-Term Confinement

Preparing psychologically for long-term confinement in a fallout shelter is essential for maintaining mental health and ensuring overall well-being during an extended period of isolation. The unique challenges of being confined to a small space with limited social interaction, resources, and external stimuli can significantly impact mental health. This chapter will explore strategies for psychological preparation, coping mechanisms, and maintaining mental resilience while living in a fallout shelter.

Understanding Psychological Challenges

Isolation and Confinement: Extended confinement can lead to feelings of isolation, loneliness, and boredom. Being cut off from normal social interactions and daily routines can affect mental health, potentially leading to anxiety, depression, and a sense of helplessness.

Limited Space and Resources: The restricted environment of a fallout shelter can create stress and frustration. Limited space, minimal privacy, and constrained resources may exacerbate feelings of discomfort and claustrophobia.

Lack of External Stimuli: A lack of exposure to natural light, fresh air, and external stimuli can impact mood and cognitive function. The absence of normal environmental changes can contribute to a sense of monotony and stagnation.

Conflict and Tension: Close quarters and prolonged interactions with others can lead to interpersonal conflicts and heightened stress. Differences in personalities, coping strategies, and expectations may cause tension within the confined group.

Mental Resilience and Adaptation

Developing a Positive Mindset:

Focus on Goals: Set short-term and long-term goals to maintain a sense of purpose and direction. These goals can include daily tasks, skill development, or personal projects. Achieving these goals can provide a sense of accomplishment and motivation.

Embrace Routine: Establish a daily routine to create structure and stability. Regular schedules for meals, exercise, work, and leisure activities help maintain a sense of normalcy and can reduce feelings of chaos or aimlessness.

Practice Gratitude: Cultivate a habit of gratitude by regularly acknowledging positive aspects of your situation. Keeping a gratitude journal or sharing positive thoughts with others can help shift focus away from stress and foster a more optimistic outlook.

Stress Management Techniques:

Relaxation Exercises: Incorporate relaxation techniques such as deep breathing, progressive muscle relaxation, or meditation into your daily routine. These practices can help reduce stress, improve sleep quality, and enhance overall mental well-being.

Physical Activity: Engage in regular physical exercise to release endorphins, improve mood, and alleviate anxiety. Create an exercise regimen that fits within the constraints of your shelter, such as bodyweight exercises, yoga, or stretching.

Creative Outlets: Pursue creative activities like drawing, writing, or crafting to channel emotions and alleviate boredom. Creative expression can be therapeutic and provide a sense of achievement and self-expression.

Building and Maintaining Relationships

Effective Communication:

Open Dialogue: Encourage open and honest communication with fellow occupants to address issues, express feelings, and resolve conflicts. Regularly checking in with each other can strengthen relationships and foster a supportive environment.

Active Listening: Practice active listening by paying attention to others' concerns and emotions. Show empathy and validation to build trust and reduce misunderstandings.

Conflict Resolution:

Problem-Solving: Address conflicts calmly and constructively by focusing on solutions rather than blame. Collaborate with others to find compromises and resolve issues in a way that respects everyone's needs and perspectives.

Time-Outs: If conflicts become intense, take time-outs to cool down and regain composure. Stepping away from the situation temporarily can help prevent escalation and allow for more thoughtful problem-solving.

Social Interaction:

Group Activities: Plan and participate in group activities to strengthen social bonds and create shared experiences. Activities such as games, group discussions, or collaborative projects can foster teamwork and camaraderie.

Personal Space: Respect each other's need for personal space and privacy. Establish boundaries to ensure that everyone has time and space to recharge and maintain their mental health.

Mental Health Maintenance and Support

Seeking Support:

Mental Health Resources: If possible, access mental health resources such as self-help books, educational materials, or virtual counseling services. These resources can provide guidance, support, and coping strategies.

Peer Support: Reach out to fellow occupants for support and encouragement. Sharing experiences, challenges, and coping strategies can create a sense of community and mutual understanding.

Monitoring Mental Health:

Self-Assessment: Regularly assess your mental health by monitoring mood, stress levels, and coping effectiveness. Be aware of signs of mental health issues such as persistent anxiety, depression, or difficulty functioning.

Professional Help: If available, seek professional help for mental health concerns. Even in a confined environment, professional guidance can be crucial for managing severe psychological issues and maintaining overall well-being.

Preparing for Psychological Challenges Before Confinement

Education and Training:

Mental Health Education: Educate yourself about psychological challenges associated with confinement and isolation. Understanding these issues in advance can help you prepare and develop effective coping strategies.

Skill Development: Develop skills such as mindfulness, emotional regulation, and problem-solving before entering the shelter. These skills can enhance resilience and improve your ability to cope with the stresses of confinement.

Building Resilience:

Pre-Shelter Training: Engage in activities that build mental resilience and adaptability. Techniques such as stress management training, emotional intelligence exercises, and coping strategy workshops can be beneficial.

Social Support Network: Establish a strong social support network outside the shelter. Having connections and support systems in place before confinement can provide a sense of stability and reassurance.

Psychological preparation for long-term confinement involves understanding potential challenges, developing coping strategies, and maintaining mental resilience. By focusing on building a positive mindset, managing stress, fostering healthy relationships, and preparing in advance, you can enhance your ability to cope with the unique stresses of living in a fallout shelter. Maintaining mental health and well-being is crucial for surviving and thriving in a confined environment, ensuring that you and your fellow occupants can navigate the challenges of long-term isolation with strength and adaptability.

Creating a Comfortable Living Space in Your Shelter

Creating a comfortable living space in your fallout shelter is essential for ensuring the well-being of its occupants during an extended stay. While the primary purpose of a shelter is to provide protection, it's equally important to make the environment as livable and comforting as possible. This chapter will guide you through the steps of designing and organizing a shelter space that promotes physical comfort, mental well-being, and a sense of normalcy, even in challenging circumstances.

Maximizing Space Efficiency

Space Planning:

Zoning: Divide the shelter into different zones or areas dedicated to specific activities, such as sleeping, eating, working, and relaxing. Clear zoning helps create a sense of order and makes the shelter feel more spacious and functional.

Multifunctional Areas: Use furniture and spaces that serve multiple purposes. For example, a dining table can also function as a workspace, or storage bins can double as seating. Multifunctional design helps maximize the use of limited space.

Storage Solutions:

Vertical Storage: Utilize vertical space by installing shelves, hooks, and racks on walls to store supplies, equipment, and personal items. Vertical storage helps free up floor space and keeps the shelter organized.

Under-Bed Storage: Make use of the space under beds or cots by storing items in bins or drawers. This area is ideal for storing less frequently used items, such as spare clothing or extra blankets.

Modular Storage Units: Consider modular storage units that can be reconfigured as needed. These units allow for flexible organization and can be adapted to changing needs over time.

Decluttering:

Minimalism: Embrace a minimalist approach to reduce clutter and keep the space tidy. Only bring essential items into the shelter, and regularly assess and remove items that are no longer needed.

Regular Organization: Set aside time for regular organization and cleaning to maintain a clutter-free environment. Keeping the shelter clean and organized contributes to a more comfortable and relaxing atmosphere.

Enhancing Physical Comfort

Sleeping Arrangements:

Comfortable Bedding: Invest in high-quality bedding, such as memory foam mattresses, thick sleeping pads, or comfortable cots. Ensure that each occupant has sufficient blankets, pillows, and sleeping bags to stay warm and comfortable.

Privacy Partitions: Create privacy partitions between sleeping areas using curtains, room dividers, or portable screens. Privacy helps occupants relax and sleep better, especially in a shared space.

Lighting:

Layered Lighting: Use a combination of overhead lighting, task lighting, and ambient lighting to create a well-lit and inviting environment. LED lights are energy-efficient and ideal for use in a shelter.

Natural Light Simulation: If the shelter lacks natural light, consider using light therapy lamps or daylight bulbs that simulate natural sunlight. These lights can help regulate circadian rhythms and improve mood.

Dimmer Switches: Install dimmer switches to adjust the brightness of lights as needed. Dimmed lighting can create a calming atmosphere in the evening, aiding in relaxation and sleep.

Temperature Control:

Insulation: Ensure that the shelter is well-insulated to maintain a stable temperature. Use insulation materials on walls, floors, and ceilings to prevent heat loss in the winter and keep the shelter cool in the summer.

Heating and Cooling: Equip the shelter with portable heaters, fans, or a climate control system to manage temperature extremes. Ensure that any heating devices are safe for indoor use and have adequate ventilation.

Layered Clothing: Encourage occupants to dress in layers to easily adjust to temperature changes. Keep extra blankets and thermal clothing available for colder conditions.

Air Quality:

Ventilation: Ensure proper ventilation to maintain good air quality. Ventilation systems should provide a continuous supply of fresh air and remove stale air, odors, and excess humidity.

Air Purifiers: Use air purifiers with HEPA filters to remove dust, allergens, and airborne particles from the shelter. This is particularly important if the shelter is sealed or has limited ventilation.

Humidity Control: Maintain a balanced humidity level (ideally between 30% and 50%) using dehumidifiers or humidifiers as needed. Proper humidity control prevents mold growth and keeps the air comfortable to breathe.

Creating a Sense of Home

Personalization:

Decorating: Allow each occupant to personalize their space with photos, artwork, or small personal items. Familiar decorations help create a sense of home and provide comfort during stressful times.

Family Mementos: Bring along family mementos, such as keepsakes, letters, or heirlooms, to remind occupants of loved ones and life outside the shelter. These items can provide emotional support and a connection to normalcy.

Comfort Items:

Soft Furnishings: Incorporate soft furnishings like cushions, rugs, and throws to make the space more inviting and cozy. These items add warmth and comfort to the shelter environment.

Entertainment: Provide entertainment options such as books, board games, puzzles, or a portable DVD player. Engaging in leisure activities helps pass the time and reduces stress.

Calming Scents: Use essential oils, scented candles (if safe), or air fresheners to introduce calming scents like lavender or chamomile. Aromatherapy can have a soothing effect and enhance the overall ambiance of the shelter.

Noise Management:

Soundproofing: Consider soundproofing the shelter to reduce external noise and create a quieter environment. Use materials like foam panels, heavy curtains, or carpets to absorb sound.

White Noise Machines: Use white noise machines or apps to mask unwanted sounds and create a peaceful atmosphere. White noise can also help improve sleep quality by drowning out background noise.

Routine and Rituals:

Daily Rituals: Establish daily rituals, such as morning coffee, evening prayers, or family meals, to create a sense of routine and normalcy. These rituals provide structure and can be comforting during uncertain times.

Meal Preparation: Make mealtime a communal activity where everyone participates in cooking, setting the table, or cleaning up. Shared meals create opportunities for bonding and maintaining a sense of family unity.

Maintaining Mental and Emotional Well-Being

Stress-Relief Strategies:

Mindfulness and Meditation: Incorporate mindfulness practices and meditation into daily routines to help occupants manage stress and stay centered. Guided meditation apps or breathing exercises can be helpful tools.

Exercise and Movement: Encourage regular physical activity, even in a confined space. Simple exercises like stretching, yoga, or bodyweight workouts can boost mood and reduce tension.

Journaling: Provide journals or notebooks for occupants to express their thoughts, feelings, and experiences. Writing can be a therapeutic outlet for processing emotions and maintaining mental clarity.

Social Interaction:

Group Activities: Plan regular group activities to foster social interaction and strengthen bonds among occupants. Activities like movie nights, group games, or collaborative projects can provide entertainment and build camaraderie.

Communication: Encourage open communication and regular check-ins to ensure that everyone feels heard and supported. Maintaining strong communication channels helps prevent misunderstandings and resolve conflicts quickly.

Mental Health Resources:

Self-Help Books: Stock the shelter with self-help books, guides on coping with stress, or resources on mental health. These materials can provide valuable advice and support during difficult times.

Access to Counseling: If possible, arrange for virtual counseling sessions or teletherapy for occupants who may need additional mental health support. Professional guidance can be critical for managing anxiety, depression, or other mental health challenges.

Preparing the Shelter for Comfort before Confinement

Shelter Setup:

Pre-Positioning Supplies: Ensure that all necessary supplies, furnishings, and comfort items are pre-positioned in the shelter before confinement begins. This includes bedding, entertainment options, and personal items.

Testing Systems: Test all shelter systems, such as lighting, climate control, and ventilation, to ensure they are functioning properly. Make any necessary adjustments or repairs before the shelter is occupied.

Occupant Orientation:

Familiarization: Allow occupants to familiarize themselves with the shelter layout and amenities before confinement. This can help reduce anxiety and make the transition to living in the shelter smoother.

Role Assignments: Assign roles or responsibilities to each occupant to give them a sense of purpose and involvement in maintaining the shelter environment. Roles can include tasks like cooking, cleaning, or managing supplies.

Creating a comfortable living space in your fallout shelter is key to ensuring the physical and emotional well-being of its occupants during an extended stay. By maximizing space efficiency, enhancing physical comfort, and fostering a sense of home, you can transform the shelter into a liveable and inviting environment.

Attending to the mental and emotional needs of occupants through stress-relief strategies, social interaction, and mental health resources is equally important for maintaining morale and resilience. With thoughtful preparation and attention to detail, your shelter can become a secure and comfortable refuge, capable of sustaining both body and spirit during even the most challenging circumstances.

Family Preparedness: Roles and Responsibilities

Family preparedness is crucial for ensuring that all members of a household are ready to respond effectively during a long-term stay in a fallout shelter. Assigning specific roles and responsibilities to each family member not only enhances the efficiency of shelter operations but also helps maintain order, reduce stress, and foster a sense of purpose and teamwork. This chapter will explore how to prepare as a family, assign roles, and establish routines that contribute to a well-organized and supportive environment during a crisis.

The Importance of Family Preparedness

Enhanced Safety and Efficiency: When every family member knows their role and responsibilities, it increases the overall safety and efficiency of the shelter. Each person contributes to the functioning of the shelter, ensuring that tasks are completed and that everyone is aware of what needs to be done.

Reduced Stress and Anxiety: Knowing what to expect and having clear duties can reduce stress and anxiety for all family members. A structured environment with defined roles helps create a sense of control and normalcy, even in difficult circumstances.

Promoting Unity and Cooperation: Assigning roles encourages cooperation and teamwork. When everyone plays a part in the shelter's operations, it fosters a sense of unity and shared responsibility, which is essential for maintaining morale during a prolonged stay.

Preparing as a Family

Family Meetings and Discussions:

Open Communication: Before an emergency occurs, hold regular family meetings to discuss potential scenarios, shelter plans, and individual responsibilities. Ensure that everyone understands the importance of their role in maintaining the shelter.

Practice Drills: Conduct practice drills to simulate emergency situations, such as a sudden need to enter the shelter or a specific task that needs to be completed quickly. Drills help reinforce what to do and ensure that everyone is prepared to act swiftly and effectively.

Assigning Roles Based on Skills and Interests:

Skill Assessment: Identify the strengths, skills, and interests of each family member. Assign roles that align with their abilities, such as first aid, cooking, communication, or maintenance. This not only ensures that tasks are completed efficiently but also helps individuals feel confident in their contributions.

Role Flexibility: While primary roles should be assigned based on strengths, ensure that all family members are cross-trained in multiple tasks. This flexibility is crucial in case someone is unable to fulfill their role due to illness or other circumstances.

Creating a Family Emergency Plan:

Written Plan: Develop a written emergency plan that outlines each person's responsibilities, the layout of the shelter, and procedures for various scenarios. Include contact information, medical details, and any other important information.

Review and Update: Regularly review and update the family emergency plan to reflect any changes in family members' abilities, needs, or shelter conditions. Ensure that everyone is familiar with the most current version of the plan.

Roles and Responsibilities

Shelter Leader:

Decision-Making: The shelter leader is responsible for making key decisions regarding the operation of the shelter, resource management, and safety. This role is typically assigned to an adult or the most experienced family member.

Coordination: The leader coordinates tasks, ensures that everyone is fulfilling their responsibilities, and addresses any issues or conflicts that arise. They also oversee the overall well-being of the family.

Medical Officer:

First Aid and Medical Care: The medical officer handles all health-related tasks, including administering first aid, managing medications, and monitoring the health of shelter occupants. They should be trained in basic first aid and have access to medical supplies.

Health Records: Maintain a log of each family member's health status, medications, and any medical incidents that occur. This log should be kept updated and accessible.

Communications Officer:

Maintaining Communication: The communications officer is responsible for operating and maintaining communication devices, such as radios, satellite phones, or ham radios. They monitor external communication channels for updates and relay important information to the family.

Daily Check-Ins: Conduct regular check-ins with external contacts, such as extended family, neighbors, or emergency services, to stay informed about the situation outside the shelter.

Supply Manager:

Inventory Management: The supply manager oversees the inventory of food, water, and other essential supplies. They track usage, manage stock levels, and ensure that resources are rationed appropriately to last the duration of the shelter stay.

Restocking and Rotation: Responsible for rotating supplies to prevent spoilage and restocking items as needed. They should also manage the organization of the shelter's storage areas.

Cook/Chef:

Meal Preparation: The cook is in charge of preparing meals for the family, ensuring that meals are nutritious, balanced, and use resources efficiently. They should be familiar with the food storage inventory and plan meals accordingly.

Kitchen Safety and Cleanliness: Maintain a clean and safe cooking area, ensuring that food is prepared in a sanitary manner. They should also manage the proper disposal of food waste to avoid attracting pests.

Sanitation Officer:

Waste Management: The sanitation officer is responsible for managing human waste, garbage, and cleaning tasks. They ensure that waste is disposed of properly and that the shelter remains clean and hygienic.

Pest Control: Monitor the shelter for signs of pests and take preventive measures to avoid infestations. They should also manage the cleaning and maintenance of waste disposal systems.

Maintenance and Safety Officer:

Shelter Maintenance: The maintenance officer handles the upkeep of the shelter, including repairs, checking ventilation systems, and ensuring that all equipment is functioning properly. They also conduct regular inspections of the shelter's structure.

Safety Protocols: Oversee safety protocols, such as fire prevention, emergency evacuation plans, and monitoring for potential hazards. They should also be responsible for managing and maintaining emergency equipment like fire extinguishers and first aid kits.

Children's Roles:

Age-Appropriate Tasks: Assign age-appropriate tasks to children to involve them in the shelter's operations and give them a sense of responsibility. Simple tasks like sorting supplies, helping with meal preparation, or assisting with cleaning can be valuable contributions.

Education and Play: Designate time for educational activities and play to ensure that children's developmental needs are met. Older children can help younger siblings with learning or lead games and activities.

Establishing Routines and Schedules

Daily Schedule:

Structured Routine: Establish a daily schedule that includes time for work, meals, exercise, leisure, and rest. A structured routine provides a sense of normalcy and helps manage time effectively.

Flexibility: While routines are important, allow for flexibility to accommodate unforeseen events or changes in circumstances. Flexibility helps prevent frustration and maintains morale.

Task Rotation:

Sharing Responsibilities: Rotate tasks among family members periodically to prevent burnout and ensure that everyone is familiar with multiple roles. Task rotation also promotes teamwork and keeps everyone engaged.

Breaks and Downtime: Ensure that all family members have regular breaks and downtime to relax and recharge. Overwork can lead to stress and fatigue, so it's important to balance responsibilities with rest.

Conflict Resolution and Support

Addressing Conflicts:

Open Communication: Encourage open communication to address conflicts as they arise. Create an environment where family members feel comfortable expressing concerns or frustrations.

Mediation: The shelter leader or a neutral party can act as a mediator to help resolve conflicts. Focus on finding solutions that meet everyone's needs and maintaining harmony in the shelter.

Emotional Support:

Mutual Support: Foster a supportive environment where family members can rely on each other for emotional support. Regularly check in on each other's well-being and offer encouragement and empathy.

Coping Strategies: Teach and practice coping strategies, such as deep breathing, mindfulness, or talking through feelings, to manage stress and emotions during challenging times.

Preparing for Emergencies within the Shelter

Emergency Drills:

Role-Specific Drills: Conduct emergency drills that focus on role-specific scenarios, such as a medical emergency, communication failure, or equipment malfunction. Drills help ensure that everyone knows how to respond quickly and effectively.

Shelter Evacuation: Practice evacuation procedures in case the shelter becomes unsafe or needs to be exited quickly. Ensure that everyone knows the evacuation plan and their specific role in executing it.

Backup Plans:

Secondary Roles: Assign secondary roles to each family member in case someone is unable to fulfill their primary responsibility. Having backup plans ensures that critical tasks are always covered.

Redundancy in Systems: Build redundancy into essential systems, such as communication, power, and sanitation, to provide backups in case of failure. This can include having spare equipment, alternative methods, or additional training.

Family preparedness is a key component of effective fallout shelter management. By assigning roles and responsibilities, creating structured routines, and fostering a supportive environment, families can work together to ensure safety, efficiency, and well-being during an extended stay in a shelter.

Preparing as a family involves not only logistical planning but also emotional readiness, communication, and mutual support. With clear roles, regular practice, and a strong sense of teamwork, families can navigate the challenges of long-term confinement with resilience and unity, turning the shelter experience into one of shared responsibility and cooperation.

Emergency Drills: Practice Makes Perfect

Emergency drills are a vital aspect of preparedness in a fallout shelter, ensuring that every family member knows how to respond effectively in various crisis situations. Regular practice through drills helps reinforce roles and responsibilities, builds confidence, and identifies potential weaknesses in your shelter's emergency plans. This chapter will guide you through the process of planning, executing, and evaluating emergency drills to ensure that you and your family are ready to handle any situation that may arise during a long-term shelter stay.

The Importance of Emergency Drills

Reinforcement of Procedures: Drills help reinforce the emergency procedures outlined in your shelter plan, ensuring that each family member is familiar with their specific roles and knows exactly what to do in different scenarios.

Building Confidence: Practicing emergency responses in a controlled environment builds confidence and reduces panic during actual emergencies. When family members know what to expect and how to react, they are more likely to remain calm and focused.

Identifying Weaknesses: Drills allow you to identify potential weaknesses or gaps in your emergency plans, equipment, or communication systems. By recognizing these issues during practice, you can address them before they become critical during a real emergency.

Promoting Teamwork: Regular drills encourage teamwork and coordination among shelter occupants. Working together to solve problems and execute procedures strengthens bonds and improves overall group efficiency.

Planning Emergency Drills

Identify Key Scenarios:

Common Scenarios: Identify the most likely emergency scenarios that could occur in your shelter, such as power outages, medical emergencies, communication failures, fire hazards, or contamination events. Focus on these scenarios for your drills.

Worst-Case Scenarios: Also consider worst-case scenarios, such as a breach in the shelter's integrity, a natural disaster impacting the shelter, or the need to evacuate. Preparing for the worst helps ensure readiness for any situation.

Set Clear Objectives:

Drill Goals: Define specific objectives for each drill, such as testing the response time for evacuation, ensuring that all emergency supplies are accessible, or practicing the use of first aid equipment. Clear objectives help measure the effectiveness of the drill.

Role-Specific Training: Tailor drills to the roles and responsibilities of each family member. For example, the medical officer might focus on first aid drills, while the communications officer practices maintaining contact with external sources.

Scheduling Drills:

Regular Practice: Schedule regular drills to maintain preparedness. Drills should be conducted frequently enough to keep procedures fresh in everyone's mind, but not so often that they become routine and lose their impact.

Variety and Surprise: Vary the scenarios and timing of drills to keep everyone on their toes. Occasionally, conduct surprise drills to simulate the unpredictability of real emergencies.

Preparation and Resources:

Resource Checklist: Before conducting a drill, ensure that all necessary resources and equipment are available and in working order. This includes emergency kits, communication devices, first aid supplies, and any specific tools needed for the scenario.

Instructions and Briefing: Provide clear instructions and a briefing before starting the drill. Make sure everyone understands the scenario, the objectives, and their roles during the drill.

Conducting Emergency Drills

Simulating Realistic Conditions:

Environmental Factors: Simulate realistic environmental conditions that might occur during an emergency, such as turning off the lights for a power outage drill or creating obstacles for an evacuation drill. The more realistic the conditions, the better prepared your family will be.

Time Constraints: Introduce time constraints to add urgency to the drill. For example, simulate a scenario where the shelter must be evacuated within a specific time frame or a medical emergency where immediate action is required.

Active Participation:

Engage All Family Members: Ensure that all family members actively participate in the drill, even if their role is secondary. Participation helps everyone stay engaged and reinforces their understanding of the procedures.

Role Rotation: Occasionally rotate roles during drills to ensure that each family member is familiar with multiple responsibilities. This cross-training can be invaluable if someone is unable to fulfill their primary role during an actual emergency.

Problem-Solving and Adaptation:

Unexpected Challenges: Introduce unexpected challenges or complications during the drill, such as a blocked exit or a malfunctioning piece of equipment. Encourage family members to problem-solve and adapt to these challenges in real-time.

Decision-Making: Emphasize the importance of quick decision-making during drills. In some scenarios, the shelter leader may need to make rapid decisions that affect the entire group, so practicing these decisions is crucial.

Evaluating and Debriefing

Post-Drill Evaluation:

Debriefing Session: After the drill, hold a debriefing session where all participants can discuss their experiences, challenges faced, and what they learned. This open discussion helps reinforce the lessons of the drill and provides valuable feedback.

Identify Strengths and Weaknesses: Evaluate the effectiveness of the drill by identifying what worked well and what needs improvement. Pay attention to response times, communication effectiveness, and the functionality of equipment.

Actionable Feedback:

Implementing Changes: Use the feedback from the drill to make necessary changes to your emergency plans, procedures, or equipment. Address any weaknesses or issues identified during the drill to improve overall preparedness.

Ongoing Improvement: Continuously improve your emergency preparedness by incorporating lessons learned from each drill. Regularly update your emergency plans based on new insights, changing conditions, or evolving needs.

Record Keeping:

Documenting Drills: Keep a record of each drill, including the date, scenario, participants, objectives, and outcomes. Documenting drills helps track progress over time and provides a reference for future training.

Updating the Emergency Plan: Based on the outcomes of the drills, update your family emergency plan to reflect any changes or new procedures. Ensure that all family members are aware of these updates and understand their roles.

Specific Drill Scenarios

Evacuation Drill:

Objective: Practice the safe and efficient evacuation of the shelter in the event of an emergency that requires leaving the shelter, such as a fire or structural damage.

Key Actions: Ensure that all family members know the evacuation routes, the location of emergency exits, and the assembly point outside the shelter. Test the functionality of emergency lighting and exit signs.

Medical Emergency Drill:

Objective: Test the response to a medical emergency, such as a severe injury, heart attack, or allergic reaction.

Key Actions: The medical officer should perform first aid or administer necessary treatment, while the communications officer contacts emergency services if possible. Ensure that first aid supplies are easily accessible and that everyone knows how to assist.

Power Outage Drill:

Objective: Simulate a power outage to practice maintaining essential functions, such as lighting, ventilation, and communication.

Key Actions: Test backup power systems, such as generators or battery backups, and ensure that everyone knows how to operate them. Practice conserving energy and managing resources during the outage.

Communication Failure Drill:

Objective: Practice maintaining communication during a scenario where primary communication systems fail, such as a radio malfunction or loss of satellite phone signal.

Key Actions: Test alternative communication methods, such as backup radios or written messages, and ensure that all family members know how to use them. Practice relaying important information without electronic devices.

Fire Safety Drill:

Objective: Prepare for a fire emergency within the shelter, focusing on fire prevention, detection, and evacuation.

Key Actions: Test smoke detectors and fire extinguishers, and practice using them. Ensure that all family members know how to evacuate the shelter safely and where to meet outside. Test the shelter's ventilation system to prevent smoke inhalation.

Preparing for Long-Term Emergencies

Extended Stay Drill:

Objective: Simulate a long-term shelter stay to practice resource management, maintaining morale, and addressing the challenges of extended confinement.

Key Actions: Test rationing food and water, rotating roles, and managing waste over an extended period. Focus on maintaining mental health and well-being through activities, communication, and routine.

Redundancy Drill:

Objective: Test the shelter's redundancy systems to ensure that backup options are functional and can be activated if needed.

Key Actions: Simulate the failure of critical systems, such as power, ventilation, or sanitation, and practice switching to backup systems. Ensure that everyone is familiar with the location and operation of redundant systems.

Emergency drills are an essential component of family preparedness in a fallout shelter. Regular practice through realistic scenarios builds confidence, reinforces roles and procedures, and ensures that everyone is ready to respond effectively in an actual emergency. By planning, conducting, and evaluating drills with clear objectives and active participation, you can identify and address potential weaknesses, improving overall safety and efficiency. With a well-practiced emergency plan and the ability to adapt to various situations, your family will be better prepared to face the challenges of long-term shelter life, knowing that they can handle any emergency with confidence and teamwork.

Stockpiling Medical Supplies: What You'll Need

Stockpiling medical supplies is a critical component of preparedness for long-term shelter living. In a fallout shelter, access to medical care may be limited or nonexistent, making it essential to have the necessary supplies to address a wide range of potential health issues. This chapter will guide you through the process of identifying, acquiring, and organizing medical supplies to ensure that you are well-equipped to handle medical emergencies, routine health needs, and the challenges of prolonged confinement.

Assessing Medical Needs

Individual Health Considerations:

Chronic Conditions: Start by assessing the specific health needs of each shelter occupant. This includes any chronic conditions such as diabetes, asthma, hypertension, or heart disease. Ensure that you stockpile medications, monitoring equipment, and supplies necessary to manage these conditions.

Allergies and Sensitivities: Take note of any allergies or sensitivities to medications, food, or environmental factors. Stock up on antihistamines, epinephrine auto-injectors (EpiPens), and other necessary treatments for allergic reactions.

Common Ailments:

Infections: Stockpile antibiotics, antifungals, and antiviral medications to treat common infections. These may include over-the-counter options as well as prescribed medications for more serious infections.

Injuries: Be prepared for injuries such as cuts, burns, sprains, and fractures. Ensure that you have an adequate supply of bandages, splints, antiseptics, and pain relievers.

Gastrointestinal Issues: Stock medications for treating common gastrointestinal problems such as nausea, diarrhoea, constipation, and indigestion. Include antacids, anti-diarrheal medications, and electrolyte solutions.

Essential Medical Supplies

First Aid Kit:

Basic Supplies: Every shelter should have a well-stocked first aid kit that includes adhesive bandages, gauze pads, medical tape, antiseptic wipes, tweezers, scissors, and a digital thermometer. These basic supplies are essential for treating minor injuries and preventing infection.

Advanced Supplies: Consider adding more advanced supplies such as sutures, hemostatic agents (to control bleeding), a CPR mask, and a blood pressure cuff. These items are particularly important if someone in the shelter has medical training.

Prescription Medications:

Extended Supply: Work with your healthcare provider to obtain an extended supply of prescription medications for each family member. Aim for at least a 30-day supply, but ideally, stockpile enough for several months.

Storage and Rotation: Store medications in a cool, dry place and regularly check expiration dates. Rotate your stock to ensure that you are using the oldest medications first and replenishing with fresh supplies.

Over-the-Counter (OTC) Medications:

Pain Relievers: Stockpile a variety of pain relievers such as acetaminophen, ibuprofen, and aspirin. These are essential for managing pain, inflammation, and fever.

Cold and Flu Medications: Include decongestants, cough suppressants, throat lozenges, and fever reducers in your stockpile to treat symptoms of colds and flu.

Digestive Aids: Antacids, anti-diarrheal medications, laxatives, and anti-nausea medications are important for managing digestive issues that may arise during long-term confinement.

Wound Care Supplies:

Bandages and Dressings: Stock a variety of bandages, including adhesive bandages, gauze pads, and elastic bandages for wrapping sprains. Include sterile dressings for covering larger wounds.

Antiseptics: Have a supply of antiseptics such as hydrogen peroxide, iodine, alcohol wipes, and antibiotic ointment to clean wounds and prevent infection.

Sterile Gloves: Include a supply of sterile gloves to protect both the patient and the caregiver during wound care or when handling medical supplies.

Medical Equipment:

Blood Pressure Monitor: A digital blood pressure monitor is essential for tracking the health of individuals with hypertension or heart conditions.

Glucose Monitor: For diabetics, a glucose monitor and a supply of test strips and lancets are crucial for managing blood sugar levels.

Thermometer: A reliable digital thermometer is necessary for monitoring fever and assessing illness.

Nebulizer/Inhaler: If anyone in the shelter has asthma or respiratory issues, ensure that you have a nebulizer or inhaler with a sufficient supply of medication.

Sanitation and Hygiene Supplies:

Hand Sanitizer: Stock up on alcohol-based hand sanitizers to maintain hygiene when soap and water are not readily available.

Soap and Disinfectants: Ensure that you have an ample supply of soap, disinfectants, and cleaning wipes to maintain cleanliness in the shelter.

Masks and Gloves: Keep a supply of disposable masks and gloves to prevent the spread of illness, especially during cold and flu season or in the event of a contagious disease.

Specialized Medical Supplies

Respiratory Supplies:

Oxygen Supply: If anyone in the shelter requires supplemental oxygen, ensure that you have a supply of oxygen tanks or concentrators. Additionally, include oxygen masks, tubing, and a pulse oximeter to monitor oxygen levels.

Nebulizer Supplies: Stock extra nebulizer kits, including mouthpieces, masks, and tubing, for individuals with respiratory conditions.

Mobility Aids:

Crutches and Braces: Have crutches, knee braces, or wrist splints available for managing sprains, fractures, or mobility issues.

Wheelchair or Walker: If someone in the shelter requires assistance with mobility, ensure that a wheelchair, walker, or cane is available and in good condition.

Emergency Medical Devices:

Defibrillator (AED): If possible, include an automated external defibrillator (AED) in your medical stockpile, especially if someone in the shelter is at risk for cardiac events.

EpiPen: For individuals with severe allergies, have a supply of epinephrine auto-injectors (EpiPens) on hand and ensure that all family members know how to use them.

Organizing and Storing Medical Supplies

Storage Solutions:

Medical Cabinet or Kit: Use a designated medical cabinet or kit to store all medical supplies in an organized manner. Ensure that the storage location is easily accessible but secure from young children.

Labeling: Clearly label all shelves, bins, and containers to make it easy to find supplies quickly in an emergency. Include expiration dates on labels to facilitate rotation of stock.

Temperature Control:

Climate Control: Store medications in a temperature-controlled environment to prevent degradation. Avoid storing medications in areas prone to temperature fluctuations, such as attics or basements.

Refrigeration: If certain medications require refrigeration, ensure that your shelter is equipped with a reliable refrigeration system, or have a plan for maintaining a cold chain during a power outage (e.g., a cooler with ice packs).

Emergency Access:

Accessibility: Ensure that essential medical supplies are easily accessible in an emergency. Consider creating a grab-and-go first aid kit that can be quickly accessed and taken with you if evacuation is necessary.

Inventory Management: Maintain an inventory list of all medical supplies, including quantities, expiration dates, and storage locations. Regularly update the inventory and review it to identify any items that need to be replenished.

Training and Education

First Aid Training:

CPR and Basic First Aid: Ensure that all shelter occupants receive training in CPR and basic first aid. Knowing how to perform these skills can be lifesaving in an emergency.

Advanced Training: If possible, take advanced first aid courses or wilderness medicine training to better prepare for medical emergencies in a confined environment.

Medication Management:

Understanding Dosages: Educate all family members on the correct dosages and administration methods for common medications. Keep a dosage chart in the medical kit for reference.

Medication Safety: Teach everyone about the importance of medication safety, including how to properly store, handle, and dispose of medications.

Using Medical Equipment:

Operation Training: Ensure that everyone knows how to operate critical medical equipment, such as a blood pressure monitor, glucose meter, or defibrillator. Practice using these devices during emergency drills.

Emergency Procedures: Develop and practice emergency procedures for medical situations, such as administering an EpiPen, performing CPR, or handling a severe injury.

Maintaining Your Medical Stockpile

Regular Checks:

Expiration Dates: Regularly check the expiration dates of all medications and supplies. Replace items that are nearing their expiration date to ensure that your stockpile remains effective.

Condition of Supplies: Inspect medical supplies for signs of wear, damage, or contamination. Replace any items that are compromised or no longer in good condition.

Replenishment:

Restocking Plan: Develop a plan for restocking medical supplies, particularly prescription medications and other consumables. Keep track of what needs to be replaced and make purchases regularly to maintain your stockpile.

Rotating Stock: Rotate your stock regularly to ensure that you are using the oldest supplies first. This practice helps prevent waste and ensures that your supplies are always fresh and effective.

Adapting to Changing Needs:

Health Changes: If a family member's health needs change, adjust your medical stockpile accordingly. This might involve adding new medications, equipment, or supplies to address emerging health concerns.

Ongoing Education: Stay informed about new medical products, treatments, or best practices that could enhance your preparedness. Incorporate new knowledge and resources into your shelter's medical plan as needed.

Stockpiling medical supplies is a crucial aspect of ensuring the health and safety of everyone in your fallout shelter during an extended stay. By carefully assessing individual health needs, gathering essential and specialized medical supplies, and organizing them effectively, you can be well-prepared to handle a wide range of medical situations.

Regular training and education are key to ensuring that all family members are capable of using the medical supplies and equipment correctly and confidently. Maintaining and regularly updating your medical stockpile will help ensure that your supplies are always ready for use when needed.

In addition to being prepared for medical emergencies, it's essential to focus on preventative care and managing chronic conditions, as this will contribute to the overall well-being of shelter occupants. With a comprehensive and well-maintained medical stockpile, you can provide the necessary care to keep everyone healthy and safe during the challenging circumstances of a nuclear fallout scenario.

Radiation-Proof Clothing and Protective Gear

Radiation-proof clothing and protective gear are essential components of a fallout shelter's preparedness plan, especially if there's a need to venture outside the shelter in the aftermath of a nuclear event. Proper protective gear can significantly reduce the risk of radiation exposure, helping to safeguard the health and safety of shelter occupants. This chapter will explore the different types of radiation-proof clothing and protective gear, their uses, and how to properly select, use, and maintain them.

Understanding Radiation and the Need for Protection

Types of Radiation:

Alpha Particles: These are large and slow-moving, and they can be stopped by a sheet of paper or the outer layer of human skin. However, they are highly dangerous if inhaled, ingested, or if they come into contact with open wounds.

Beta Particles: Beta particles are smaller and faster than alpha particles and can penetrate the skin, causing radiation burns or internal damage if inhaled or ingested.

Gamma Rays and X-Rays: These are highly penetrating forms of radiation that can pass through the body and require dense materials like lead or concrete for protection.

Neutron Radiation: This type of radiation is very penetrating and difficult to shield against, requiring materials like water, concrete, or specialized neutron-absorbing materials for protection.

The Role of Protective Gear: Radiation-proof clothing and protective gear provide a critical layer of defense against radiation, particularly beta particles and gamma rays. While they cannot offer complete protection, they can reduce exposure levels and help prevent radioactive contamination on the skin and clothing.

Types of Radiation-Proof Clothing and Gear

Radiation-Proof Suits:

Hazmat Suits: Hazmat (hazardous materials) suits are full-body suits designed to protect against a variety of chemical, biological, and radiological hazards. They are often made from materials like Tyvek or PVC and can include a respirator to protect against inhalation of radioactive particles.

NBC Suits: NBC (Nuclear, Biological, Chemical) suits are specifically designed for protection against nuclear, biological, and chemical threats. They are typically made from multi-layered fabrics that block radiation and prevent contamination. These suits usually include a hood, gloves, and boots for full-body coverage.

Respiratory Protection:

Gas Masks: Gas masks with appropriate filters (such as P3 or NBC filters) protect against inhaling radioactive dust, smoke, and particles. They should be used whenever there is a risk of airborne contamination.

Full-Face Respirators: These provide more comprehensive protection than half-mask respirators by covering the entire face. They typically come with replaceable filters that can be selected based on the specific threat (e.g., radioactive particles, chemical vapors).

Powered Air-Purifying Respirators (PAPR): PAPRs provide a higher level of respiratory protection by using a battery-powered blower to push air through a filter and into a full-face mask or hood. This reduces breathing resistance and provides a constant flow of filtered air.

Protective Gloves:

Radiation-Resistant Gloves: These gloves are made from materials that block radiation, such as lead or specialized synthetic fabrics. They are essential for handling contaminated materials or when there is a risk of skin exposure to radiation.

Disposable Gloves: In addition to radiation-resistant gloves, disposable gloves can be worn underneath to provide an additional barrier against contamination. They should be changed frequently to avoid the buildup of radioactive particles.

Footwear:

Radiation-Proof Boots: Boots made from radiation-resistant materials or heavy-duty rubber can protect the feet and lower legs from radiation exposure. They often include steel toes and shanks for additional protection.

Disposable Boot Covers: These are used to prevent contamination of footwear. They can be easily removed and discarded after exposure to radioactive materials, preventing the spread of contamination.

Head and Eye Protection:

Radiation Shielding Hoods: These hoods are designed to protect the head, neck, and shoulders from radiation exposure. They are typically worn in conjunction with other protective gear like hazmat suits.

Lead-Lined Goggles or Visors: Eye protection is essential when dealing with gamma radiation, as it can penetrate the eyes and cause damage. Lead-lined goggles or visors help block radiation and protect the eyes.

Dosimeters and Radiation Detectors:

Personal Dosimeters: These small, wearable devices measure an individual's exposure to radiation over time. They provide real-time feedback, allowing you to monitor your exposure and take action if levels become dangerous.

Geiger Counters: Geiger counters are handheld devices used to detect and measure radiation in the environment. They are essential for assessing the safety of an area before entering and for ongoing monitoring during and after exposure.

Selecting the Right Protective Gear

Assessing the Risk:

Type of Radiation: Choose protective gear based on the type of radiation you are likely to encounter. For example, gamma radiation requires more robust protection, such as lead-lined materials, while alpha and beta radiation can be blocked by lighter fabrics.

Duration of Exposure: Consider how long you might need to wear the protective gear. For extended periods, comfort and ease of movement become important factors, as does the ability to manage heat and moisture buildup inside the suit.

Comfort and Fit:

Proper Sizing: Ensure that all protective gear fits properly. Ill-fitting gear can reduce effectiveness and cause discomfort, which may lead to improper use or increased risk of exposure.

Adjustability: Look for gear that is adjustable, particularly in the case of respirators, suits, and gloves. Adjustable straps and closures help create a secure fit and enhance protection.

Quality and Certification:

Certified Gear: Always choose protective gear that is certified for radiation protection by reputable organizations, such as the National Institute for Occupational Safety and Health (NIOSH) or the European Committee for Standardization (CEN).

Durability: Ensure that the gear is made from high-quality, durable materials that can withstand the rigors of use in a contaminated environment. The gear should be resistant to tears, punctures, and chemical degradation.

Proper Use and Maintenance of Protective Gear

Donning and Doffing:

Safe Procedures: Practice donning (putting on) and doffing (taking off) protective gear to ensure that you can do so safely and efficiently. Improper removal of gear can result in contamination of your skin or clothing.

Buddy System: Whenever possible, use a buddy system when donning and doffing protective gear. Having another person assist you can help ensure that all gear is properly secured and that no skin is exposed.

Decontamination:

Decontamination Protocols: After use, protective gear should be decontaminated according to established protocols. This typically involves washing with soap and water, followed by rinsing with a decontamination solution.

Disposable Gear: Use disposable items such as gloves, boot covers, and face masks whenever possible. These can be safely discarded after use, reducing the risk of spreading contamination.

Storage:

Proper Storage Conditions: Store protective gear in a clean, dry, and secure location, away from direct sunlight, chemicals, and extreme temperatures. This helps maintain the integrity of the materials and ensures they are ready for use when needed.

Regular Inspections: Regularly inspect all protective gear for signs of wear, damage, or degradation. Replace any gear that shows signs of weakness or that has reached its expiration date.

Training and Drills:

Regular Training: Ensure that all shelter occupants are trained in the proper use of radiation-proof clothing and protective gear. Training should include how to don, doff, and decontaminate the gear, as well as how to respond to different radiation exposure scenarios.

Emergency Drills: Incorporate the use of protective gear into emergency drills to ensure that everyone is familiar with the equipment and can use it effectively under stress.

Preparing for Long-Term Use

Backup Gear:

Multiple Sets: Stock multiple sets of protective gear to ensure that you have backups available in case of damage or contamination. This is particularly important for items like gloves, masks, and suits.

Size Variations: Include gear in various sizes to accommodate different body types and ensure that everyone in the shelter has access to properly fitting protection.

Long-Term Maintenance:

Regular Cleaning: Even if gear has not been used, it should be regularly cleaned and inspected to prevent deterioration. Follow manufacturer guidelines for cleaning and maintenance.

Component Replacement: Some protective gear, like respirators, may require periodic replacement of components such as filters or seals. Keep a supply of replacement parts on hand and regularly check that they are in good condition.

Radiation-proof clothing and protective gear are essential components of a comprehensive fallout shelter preparedness plan. By carefully selecting, using, and maintaining the right protective gear, you can significantly reduce the risk of radiation exposure and ensure the safety of shelter occupants if they need to venture outside or handle contaminated materials. Understanding the different types of radiation and the appropriate protective measures for each is critical. With proper training, regular practice, and diligent maintenance, your shelter will be equipped to handle the challenges of radiation exposure, allowing you to focus on survival and well-being in a secure environment.

Understanding Fallout Shelters: Historical Lessons

Understanding the history and evolution of fallout shelters provides valuable insights into their design, effectiveness, and the lessons learned from past experiences. These lessons can guide modern efforts to prepare and maintain fallout shelters that are better equipped to protect against nuclear fallout and other catastrophic events. This chapter will explore the historical development of fallout shelters, examining key events, shelter designs, and the lessons that can be applied to today's preparedness efforts.

The Origins of Fallout Shelters

Early Concepts of Sheltering:

World War II Air Raid Shelters: The concept of shelters to protect civilians from aerial bombardment was widely implemented during World War II. In cities like London, people used underground shelters, subway stations, and reinforced basements to survive bombings. These early shelters laid the groundwork for the idea of protecting against more advanced threats like nuclear fallout.

Post-War Nuclear Fears: Following the atomic bombings of Hiroshima and Nagasaki, the destructive potential of nuclear weapons became evident. This spurred the development of shelters specifically designed to protect against nuclear fallout, rather than just conventional bombings.

Cold War Era Shelters:

Civil Defense Initiatives: During the Cold War, especially in the 1950s and 1960s, the fear of nuclear war between the United States and the Soviet Union led to widespread civil defense initiatives. Governments encouraged citizens to build fallout shelters in their homes, and public shelters were constructed in cities across the world.

Public Awareness Campaigns: Governments used public awareness campaigns, such as the U.S. "Duck and Cover" drills, to educate people about the dangers of nuclear fallout and the importance of sheltering. These campaigns emphasized the need for preparedness and the role of fallout shelters in surviving a nuclear attack.

Evolution of Fallout Shelter Designs

Early Designs:

Basement Shelters: One of the earliest and simplest forms of fallout shelters involved reinforcing a section of a home's basement. Homeowners were advised to stockpile food, water, and basic medical supplies in these shelters, which were often lined with concrete or lead to reduce radiation exposure.

Underground Bunkers: More elaborate shelters were built underground, offering greater protection from both blast effects and radiation. These bunkers were designed with thick walls, ventilation systems, and secure entrances to withstand the initial shockwave of a nuclear explosion.

Public Fallout Shelters:

School and Office Shelters: Many schools, government buildings, and office complexes were equipped with designated fallout shelters, often located in basements or underground areas. These shelters were stocked with basic supplies and intended to house large groups of people for short periods.

Community Shelters: Some communities constructed larger fallout shelters capable of accommodating hundreds of people. These shelters were typically located in public buildings, such as libraries, courthouses, and subway stations, and were equipped with more extensive supplies and amenities.

Advances in Shelter Technology:

Air Filtration Systems: As understanding of radioactive fallout improved, so too did the design of fallout shelters. Modern shelters often include advanced air filtration systems capable of removing radioactive particles from the air, ensuring a supply of clean air for occupants.

Radiation Shielding Materials: Advances in materials science led to the development of more effective radiation shielding, using materials like lead, concrete, and specialized synthetic fabrics. These improvements enhanced the protective capabilities of shelters, making them more effective at blocking harmful radiation.

Key Historical Events and Lessons Learned

The Cuban Missile Crisis (1962):

Heightened Awareness: The Cuban Missile Crisis brought the threat of nuclear war to the forefront of public consciousness. During this period, there was a surge in the construction of fallout shelters, both private and public, as people feared an imminent nuclear exchange.

Importance of Preparedness: The crisis highlighted the importance of being prepared for sudden and severe threats. It underscored the need for shelters to be ready at a moment's notice, with supplies and systems in place to support extended stays.

The Chernobyl Disaster (1986):

Real-World Fallout Exposure: Although not a nuclear attack, the Chernobyl disaster provided valuable lessons about the dangers of radioactive fallout. The widespread contamination across Europe demonstrated how far and how quickly radioactive particles could spread, emphasizing the need for effective sheltering.

Long-Term Contamination: Chernobyl also taught the world about the long-term environmental and health impacts of radioactive contamination. This event reinforced the importance of proper decontamination procedures and the potential need for extended sheltering in the aftermath of a nuclear incident.

The Fukushima Daiichi Nuclear Disaster (2011):

Modern Shelter Challenges: The Fukushima disaster, caused by a tsunami following a massive earthquake, revealed challenges in sheltering from nuclear fallout in the modern era. It highlighted the importance of not only having shelters but also ensuring they are resilient to multiple types of disasters.

Preparedness and Response: The disaster showed the need for robust emergency response plans that include radiation-proof clothing, protective gear, and proper evacuation procedures. It also emphasized the importance of public awareness and education on how to respond to nuclear emergencies.

Lessons for Modern Fallout Shelter Design

Comprehensive Planning:

Integrated Systems: Modern fallout shelters should be designed with integrated systems that address all aspects of survival, including air filtration, water purification, waste management, and communication. These systems need to be robust and reliable, capable of functioning independently for extended periods.

Modularity and Scalability: Shelters should be modular and scalable, allowing for expansion or adaptation based on the number of occupants or the severity of the situation. This flexibility can be critical in responding to unexpected challenges.

Location and Accessibility:

Strategic Placement: The location of a fallout shelter is crucial. It should be easily accessible from the living area, ideally within or adjacent to the home, to allow for quick entry in an emergency. Additionally, shelters should be located in areas that minimize exposure to external hazards, such as away from floodplains or potential blast zones.

Multiple Access Points: Shelters should have multiple access points or escape routes to ensure that occupants can enter and exit safely, even if one entryway is blocked or compromised.

Durability and Resilience:

Structural Integrity: Modern shelters must be built to withstand not only nuclear blasts and radiation but also other natural disasters, such as earthquakes or floods. This requires careful consideration of building materials and construction techniques.

Maintenance and Upkeep: Regular maintenance and inspections are essential to ensure that shelters remain in good condition and that all systems are functioning properly. This includes testing ventilation, filtration, and power systems, as well as rotating supplies.

Public Awareness and Education:

Ongoing Education: One of the most significant lessons from history is the importance of public awareness and education. People need to understand the risks of nuclear fallout and how to properly use and maintain shelters. Ongoing education efforts, including drills and training, are crucial for effective preparedness.

Community Involvement: Encouraging community involvement in shelter planning and preparedness can enhance overall resilience. Communities with shared shelters or coordinated response plans are better equipped to support each other during a crisis.

Modern Fallout Shelters: A Synthesis of Historical Lessons

Advances in Technology:

Smart Shelters: Modern technology offers the potential for "smart" fallout shelters equipped with automated systems for air filtration, water purification, and radiation monitoring. These systems can be connected to external sensors and controlled remotely, providing real-time data and enhancing safety.

Renewable Energy: Incorporating renewable energy sources, such as solar panels or wind turbines, into shelter design can provide a sustainable power supply, reducing reliance on external fuel sources and enhancing long-term survivability.

Psychological Preparedness:

Mental Health Considerations: Historical experiences have shown that long-term confinement in a shelter can take a toll on mental health. Modern shelters should include features that support psychological well-being, such as comfortable living spaces, entertainment options, and opportunities for social interaction.

Stress Management: Preparing for the psychological challenges of shelter living is just as important as physical preparedness. This includes stress management training, establishing routines, and ensuring that occupants have access to mental health resources.

The history of fallout shelters offers valuable lessons that can inform modern preparedness efforts. By understanding the evolution of shelter design, the challenges faced during key historical events, and the advances in technology and materials, we can build shelters that are more effective, resilient, and comfortable. Modern fallout shelters must be comprehensive in their design, integrating lessons from the past with the latest technology and best practices. This synthesis of historical knowledge and modern innovation will ensure that shelters are equipped to protect against a wide range of threats, providing a safe haven for occupants in the event of a nuclear incident or other catastrophic events.

How Fallout Shelters Have Evolved Over Time

The evolution of fallout shelters over time reflects advancements in technology, changes in societal attitudes toward nuclear threats, and lessons learned from historical events. As our understanding of radiation and nuclear warfare has deepened, the design, construction, and purpose of fallout shelters have also evolved. This chapter will explore the key stages of this evolution, from early rudimentary shelters to the sophisticated, modern facilities designed to protect against a range of catastrophic events.

The Birth of Fallout Shelters: Early Designs and Concepts

World War II and Air Raid Shelters:

Origins of Sheltering: The concept of civilian shelters gained prominence during World War II, with air raid shelters being constructed to protect against bombings. These shelters, often hastily built, provided basic protection from blasts and debris, but they were not designed with radiation in mind.

Basements and Underground Spaces: Many early shelters were simply reinforced basements or underground spaces. The emphasis was on providing a safe place to wait out an air raid, with little consideration for long-term habitation or radiation protection.

Post-War Nuclear Fears and the Cold War:

The Dawn of the Nuclear Age: The development of nuclear weapons and the bombings of Hiroshima and Nagasaki highlighted the need for more specialized protection against the unique dangers of nuclear fallout. The early post-war period saw the first attempts to create shelters specifically designed to protect against radiation.

Government Initiatives: During the early Cold War, governments, particularly in the United States and the Soviet Union, began promoting the construction of fallout shelters. Public campaigns encouraged citizens to build their own shelters at home, often using simple designs that could be constructed with readily available materials.

Cold War Fallout Shelters: Mass Production and Standardization

The 1950s Boom in Shelter Construction:

Civil Defense Programs: The 1950s saw a massive increase in the construction of fallout shelters, driven by government civil defense programs. In the U.S., the Federal Civil Defense Administration (FCDA) provided plans and guidelines for building home fallout shelters, and many families took on the task of constructing their own.

Standardized Designs: To make shelter construction more accessible, standardized designs were developed. These often included reinforced concrete structures with thick walls to protect against radiation and were equipped with basic ventilation and sanitation systems.

Public Shelters and Community Preparedness:

Large-Scale Public Shelters: Alongside private shelters, governments began constructing public fallout shelters in schools, government buildings, and other public spaces. These shelters were intended to protect large numbers of people in the event of a nuclear attack.

Stockpiling Supplies: Public shelters were typically stocked with basic supplies such as food, water, first aid kits, and radiation detection equipment. The idea was to provide for occupants for a few days to a week until it was safe to leave or until help arrived.

Cultural Impact:

Shelters in Popular Media: The proliferation of fallout shelters had a significant cultural impact, with shelters often depicted in films, television shows, and literature of the era. These depictions ranged from optimistic portrayals of preparedness to darker, more dystopian visions of a post-nuclear world.

Technological Advancements in Shelter Design

The 1960s and Beyond: Innovation in Materials and Systems:

Advanced Materials: As understanding of radiation protection improved, so did the materials used in shelter construction. Lead-lined walls, thicker concrete barriers, and specialized coatings became more common, offering better protection against gamma rays and other forms of radiation.

Air Filtration Systems: The development of more sophisticated air filtration systems allowed shelters to better protect occupants from airborne radioactive particles. These systems, often incorporating HEPA filters and activated carbon, became standard in newer shelters.

Deeper Underground Bunkers:

Military Bunkers: Military applications of fallout shelters pushed the envelope in terms of depth and protection. Facilities like NORAD's Cheyenne Mountain Complex were built deep underground and designed to withstand direct nuclear strikes. These bunkers featured extensive systems for air filtration, water purification, and power generation.

Civilian Underground Shelters: Inspired by military designs, some civilian shelters also moved underground. These shelters offered superior protection by using the earth itself as a barrier against radiation and shockwaves. They often included reinforced blast doors, multiple layers of protection, and more advanced life-support systems.

The Decline and Resurgence of Fallout Shelters

Post-Cold War Decline:

Reduced Perceived Threat: With the end of the Cold War, the perceived threat of nuclear war diminished, leading to a decline in the construction and maintenance of fallout shelters. Many public shelters were repurposed, decommissioned, or simply forgotten.

Shifting Focus: During the 1990s and early 2000s, the focus of civil defense shifted toward other threats, such as terrorism and natural disasters. Emergency preparedness became more about general resilience rather than specific protection against nuclear fallout.

21st Century Resurgence:

Renewed Nuclear Concerns: In recent years, concerns about nuclear proliferation, geopolitical instability, and the potential for terrorist attacks involving nuclear or radiological weapons have led to a resurgence of interest in fallout shelters.

Modern Preppers and Private Shelters: The modern "prepper" movement has seen an increase in individuals and families building private shelters, often with advanced features and designed for long-term survival. These shelters are typically well-stocked and equipped with modern conveniences, reflecting lessons learned from past designs.

Modern Fallout Shelters: State-of-the-Art Design

Smart Shelters and Technology Integration:

Automation and Monitoring: Modern fallout shelters often incorporate smart technology, with automated systems for air filtration, power management, and radiation monitoring. These systems can be controlled remotely, providing real-time data and alerts.

Renewable Energy: To reduce reliance on external power sources, many modern shelters are equipped with renewable energy systems, such as solar panels or wind turbines, along with battery storage to ensure continuous operation.

Sustainable Living in Shelters:

Water and Waste Management: Advances in water purification and waste management systems allow for sustainable living in shelters for extended periods. Greywater recycling, composting toilets, and rainwater collection are commonly integrated into shelter designs.

Food Production: Some modern shelters include hydroponic or aquaponic systems for growing food, reducing dependence on stored supplies and enhancing long-term survival prospects.

Enhanced Comfort and Liveability:

Ergonomic Design: Modern shelters place a greater emphasis on comfort and livability, with ergonomic designs that maximize space efficiency and minimize the psychological stress of long-term confinement. Lighting, climate control, and noise reduction are key considerations.

Mental Health Support: Recognizing the psychological challenges of shelter living, many modern designs include features to support mental health, such as entertainment systems, exercise equipment, and spaces for relaxation and social interaction.

Lessons from the Past and Future Directions

Incorporating Historical Lessons:

Preparedness and Flexibility: One of the key lessons from the history of fallout shelters is the importance of preparedness and flexibility. Shelters should be designed to adapt to a variety of threats, not just nuclear fallout, and should be regularly updated and maintained.

Community and Collaboration: Historically, the most successful shelter initiatives involved strong community support and collaboration. Modern shelter efforts can benefit from similar approaches, encouraging shared resources and mutual aid among neighbors.

Future Trends in Shelter Design:

Hybrid Shelters: Future fallout shelters may continue to evolve into hybrid designs that combine features of traditional bunkers with modern eco-friendly and sustainable technologies. These shelters will be capable of protecting against a wide range of threats while also supporting a self-sufficient lifestyle.

Virtual Reality and Mental Health: As technology advances, virtual reality (VR) could play a role in enhancing the mental well-being of shelter occupants, providing immersive environments that reduce the sense of confinement and isolation.

Global Preparedness: The evolution of fallout shelters is likely to continue as global awareness of nuclear threats and other disasters grows. International cooperation and the sharing of best practices will be essential in developing the next generation of shelters.

The evolution of fallout shelters over time reflects a combination of technological advancements, historical lessons, and changing societal needs. From simple basements to sophisticated underground bunkers, shelters have continuously adapted to meet the challenges of the times.

Modern fallout shelters are the culmination of decades of learning and innovation, incorporating the best of past designs while embracing new technologies and sustainable practices.

Government Guidelines and Resources for Fallout Shelters

Government guidelines and resources for fallout shelters have evolved over the decades in response to the changing nature of threats and advancements in technology. These guidelines are designed to help citizens prepare for the possibility of nuclear fallout by providing standardized information on how to build, equip, and maintain fallout shelters. This chapter will explore the key government initiatives, guidelines, and resources available for individuals and communities to prepare their own fallout shelters.

Historical Context: The Birth of Government Guidelines

Cold War Era Initiatives:

Federal Civil Defense Administration (FCDA): Established in 1950, the FCDA was one of the first government bodies in the United States dedicated to civil defense. It produced extensive materials, including pamphlets, manuals, and public service announcements, to educate the public on building and stocking fallout shelters. The FCDA's efforts were part of a broader civil defense strategy aimed at preparing the population for a potential nuclear attack.

The Office of Civil Defense (OCD): In 1961, the Office of Civil Defense was established under the Department of Defense, taking over many responsibilities from the FCDA. The OCD expanded on earlier guidelines, providing more detailed instructions on constructing shelters, as well as organizing community shelters in public buildings.

Public Education Campaigns:

Duck and Cover: One of the most famous public education campaigns, "Duck and Cover," was launched in the early 1950s. It taught schoolchildren and the general public how to protect themselves during a nuclear explosion, emphasizing the importance of immediate sheltering.

Shelter Construction Manuals: During the Cold War, the U.S. government published numerous manuals and guides on how to build fallout shelters. These guides provided step-by-step instructions for constructing shelters using common building materials and offered advice on how to stock them with essential supplies.

Key Government Guidelines for Fallout Shelters

Design and Construction Standards:

Department of Defense (DoD) Standards: The DoD developed detailed guidelines for constructing fallout shelters that could withstand nuclear blasts and protect occupants from radiation. These guidelines covered everything from the thickness of walls to the type of materials to use for radiation shielding.

Federal Emergency Management Agency (FEMA) Guidelines: FEMA, which absorbed the civil defense responsibilities of earlier agencies, has continued to provide guidelines for fallout shelters. These include recommendations on shelter design, ventilation, sanitation, and the storage of food and water. FEMA's "Nuclear Attack Planning Base" (NAPB) documents offer detailed information on shelter locations and requirements based on different threat scenarios.

Community Shelter Programs:

Public Shelter Identification: During the Cold War, the U.S. government identified and marked thousands of public buildings as designated fallout shelters. These buildings were stocked with basic supplies and were intended to provide refuge for large numbers of people. The yellow and black fallout shelter signs became an iconic symbol of this era.

Community Preparedness Programs: Governments encouraged communities to organize and prepare for nuclear emergencies by creating local civil defense teams. These teams were responsible for ensuring that public shelters were maintained and that the community was informed and prepared.

Resources for Building and Stocking Fallout Shelters

Federal Emergency Management Agency (FEMA) Resources:

"In Time of Emergency" Guide: One of FEMA's key resources, this guide offers comprehensive instructions on how to prepare for a nuclear disaster, including how to build and equip a fallout shelter. It covers everything from selecting a location and materials to ensuring adequate ventilation and stocking the shelter with essential supplies.

"Are You Ready?" Guide: This all-hazards guide includes a section specifically on nuclear preparedness. It provides practical advice on what to do before, during, and after a nuclear event, including how to use a fallout shelter effectively.

Ready.gov:

Nuclear Explosion Preparedness: The Ready.gov website, maintained by FEMA, provides up-to-date information on how to prepare for a nuclear explosion. It includes guidelines on what supplies to have, how to shelter in place, and how to protect against radiation exposure.

Emergency Supply Lists: Ready.gov also offers checklists for emergency supplies, including specific recommendations for fallout shelters. These lists cover food, water, medical supplies, and other essentials needed for long-term sheltering.

Department of Energy (DOE) and National Laboratories:

Radiation Shielding Information: The DOE and its associated national laboratories have produced extensive research on radiation shielding, which has informed government guidelines on shelter construction. This research is available to the public and can help individuals design shelters that provide effective protection against radiation.

Nuclear Incident Response Teams (NIRT): The DOE's NIRT provides technical expertise and support in the event of a nuclear or radiological incident. While primarily focused on response, the team's research and resources contribute to the broader body of knowledge on nuclear safety and sheltering.

International Guidelines and Resources

International Atomic Energy Agency (IAEA):

Safety Standards: The IAEA sets international safety standards for radiation protection, including guidelines for building shelters to protect against nuclear fallout. These standards are used by governments worldwide to develop their own sheltering guidelines.

Publications and Resources: The IAEA publishes a wide range of resources on nuclear safety, radiation protection, and emergency preparedness. These resources are available to the public and can be used to inform the construction and maintenance of fallout shelters.

United Nations Office for Disaster Risk Reduction (UNDRR):

Disaster Risk Reduction Frameworks: The UNDRR provides frameworks and guidelines for reducing disaster risk, including the risk of nuclear fallout. While not specifically focused on fallout shelters, their resources emphasize the importance of preparedness and resilience, which are critical for effective sheltering.

Global Platform for Disaster Risk Reduction: This platform facilitates the sharing of best practices and knowledge among countries and communities. It includes discussions on nuclear safety and the role of shelters in disaster risk reduction.

European Union (EU) Guidelines:

Nuclear Safety and Radiation Protection: The EU has developed guidelines for member states on nuclear safety and radiation protection. These guidelines include recommendations for fallout shelters, particularly in regions near nuclear power plants or other potential sources of radioactive contamination.

Cross-Border Cooperation: The EU promotes cross-border cooperation in disaster preparedness, including nuclear safety. Member states share resources and information to improve the effectiveness of fallout shelters and other protective measures.

Modern Developments in Government Guidelines

Advances in Shelter Technology:

Incorporation of Smart Technology: Modern government guidelines increasingly emphasize the use of smart technology in fallout shelters. This includes automated systems for air filtration, radiation monitoring, and energy management, which can enhance the safety and comfort of shelter occupants.

Sustainability and Self-Sufficiency: Governments are also promoting the integration of sustainable practices in shelter design, such as renewable energy sources, water recycling, and food production systems. These features are critical for long-term sheltering scenarios.

Preparedness Campaigns and Drills:

National Preparedness Month: In the United States, September is designated as National Preparedness Month. During this time, FEMA and other agencies promote public awareness of disaster preparedness, including the importance of fallout shelters. Resources, drills, and exercises are provided to help individuals and communities prepare effectively.

International Drills and Exercises: Countries around the world conduct regular drills and exercises to test their nuclear preparedness plans. These exercises often include scenarios involving fallout shelters, helping to ensure that guidelines are practical and effective in real-world situations.

Accessibility and Inclusivity:

Guidelines for Vulnerable Populations: Modern guidelines increasingly address the needs of vulnerable populations, such as the elderly, disabled, and children, in fallout shelter planning. This includes recommendations for accessible shelter designs, special medical supplies, and tailored communication strategies.

Multilingual Resources: To ensure that information is accessible to all, governments are producing guidelines and resources in multiple languages. This is especially important in diverse communities where English may not be the primary language.

How to Access and Utilize Government Resources

Online Resources:

Official Websites: Access the latest government guidelines and resources for fallout shelters through official websites such as FEMA.gov, Ready.gov, and the IAEA's website. These sites offer downloadable guides, checklists, and planning tools.

Apps and Digital Tools: Some governments offer mobile apps that provide real-time updates, alerts, and resources for disaster preparedness, including fallout shelters. These apps can be invaluable during an emergency, offering guidance on the go.

Local Government and Community Programs:

Community Workshops and Training: Many local governments offer workshops and training programs on disaster preparedness, including how to build and stock a fallout shelter. These programs often provide hands-on guidance and connect you with local resources.

Emergency Management Offices: Contact your local emergency management office for information on regional guidelines, public shelter locations, and community preparedness initiatives. They can also provide advice on tailoring national guidelines to local conditions.

Consulting Experts:

Civil Engineers and Architects: For those building a new fallout shelter, consulting with civil engineers or architects who specialize in disaster-resistant structures can ensure that the shelter meets or exceeds government guidelines.

Emergency Preparedness Consultants: These professionals can help you assess your needs, develop a customized shelter plan, and ensure that your shelter is properly equipped and maintained.

Government guidelines and resources for fallout shelters are critical tools in preparing for nuclear emergencies. By understanding and utilizing these resources, individuals and communities can build shelters that are effective, safe, and tailored to their specific needs. From historical initiatives to modern advancements, the evolution of these guidelines reflects the ongoing effort to protect citizens from the dangers of nuclear fallout. Staying informed and proactive in shelter planning is key to ensuring that you and your loved ones are ready to face any potential nuclear threat with confidence and resilience.

The Impact of Nuclear Fallout on the Environment

Nuclear fallout has devastating and long-lasting effects on the environment. The release of radioactive materials during a nuclear explosion or accident contaminates the air, soil, water, and living organisms, leading to widespread ecological damage. Understanding the impact of nuclear fallout on the environment is essential for developing effective mitigation strategies and for preparing for the long-term consequences of a nuclear event. This chapter explores the various ways in which nuclear fallout affects the environment, the mechanisms of contamination, and the potential long-term consequences for ecosystems and human health.

Understanding Nuclear Fallout

What is Nuclear Fallout?:

Radioactive Debris: Nuclear fallout refers to the radioactive particles that are propelled into the upper atmosphere following a nuclear explosion or accident. These particles eventually fall back to Earth, contaminating the environment as they settle.

Types of Fallout: Fallout can be divided into two categories: early fallout, which occurs within the first 24 hours and is more concentrated near the site of the explosion, and delayed fallout, which occurs over days, weeks, or even months and can spread across vast areas.

Sources of Fallout:

Nuclear Explosions: Both atmospheric and underground nuclear explosions release large amounts of radioactive material. Atmospheric tests, in particular, disperse fallout over wide areas due to the high altitude at which the particles are released.

Nuclear Accidents: Accidents at nuclear power plants, such as the Chernobyl disaster in 1986 and the Fukushima Daiichi disaster in 2011, also result in significant fallout. These events release radioactive materials directly into the environment, leading to widespread contamination.

Immediate Environmental Impact

Airborne Contamination:

Radioactive Clouds: In the immediate aftermath of a nuclear explosion or accident, a radioactive cloud forms, containing particles of various sizes. These particles can be carried by wind currents over long distances, depending on the altitude and weather conditions.

Inhalation Risks: As these radioactive particles fall to the ground, they contaminate the air, posing significant health risks to both humans and animals through inhalation. Inhaled radioactive particles can lodge in the lungs, increasing the risk of cancers and other health issues.

Soil Contamination:

Deposition of Fallout: As fallout settles, it contaminates the soil with radioactive isotopes such as cesium-137, iodine-131, and strontium-90. These isotopes have varying half-lives, ranging from days to decades, meaning they remain hazardous for extended periods.

Absorption by Plants: Radioactive particles deposited on the soil can be absorbed by plants, entering the food chain. This process not only affects the plants themselves but also the animals and humans who consume them.

Water Contamination:

Surface Water: Fallout can contaminate rivers, lakes, and oceans as radioactive particles settle on the surface of the water or are carried into bodies of water by rain. This contamination poses risks to aquatic life and can make water sources unsafe for human consumption.

Groundwater: Radioactive particles can seep into the soil and contaminate groundwater supplies. This contamination is particularly concerning because groundwater is a major source of drinking water for many communities.

Wildlife Impact:

Acute Radiation Exposure: Wildlife in the immediate vicinity of a nuclear event can suffer from acute radiation sickness, leading to high mortality rates. Animals exposed to high levels of radiation may experience burns, organ failure, and genetic mutations.

Disruption of Ecosystems: The death of plants and animals due to radiation exposure can disrupt entire ecosystems, affecting species diversity and leading to long-term ecological imbalances.

Long-Term Environmental Consequences

Soil Degradation:

Persistent Contamination: Radioactive isotopes with long half-lives, such as cesium-137 and strontium-90, can persist in the soil for decades. These isotopes continuously emit radiation, posing ongoing risks to living organisms and making the land unusable for agriculture or habitation.

Soil Erosion: In some cases, the removal of vegetation due to radiation exposure can lead to increased soil erosion. This further degrades the land, making it more difficult for ecosystems to recover.

Impact on Agriculture:

Contaminated Crops: Crops grown in contaminated soil or irrigated with contaminated water can accumulate radioactive isotopes. This makes the crops unsafe for consumption and can lead to widespread food shortages.

Livestock Exposure: Livestock that grazes on contaminated grass or drinks contaminated water can also accumulate radioactive isotopes in their bodies. This poses a risk not only to the animals themselves but also to the people who consume meat, milk, and other animal products.

Aquatic Ecosystems:

Bioaccumulation: Radioactive isotopes can accumulate in the bodies of aquatic organisms, particularly in fish and other species at the top of the food chain. This bioaccumulation can lead to significant ecological damage and make seafood unsafe for consumption.

Long-Term Water Contamination: Contaminated water bodies may remain hazardous for years, as radioactive particles settle into sediments and continue to leach into the water. This long-term contamination can disrupt aquatic ecosystems and affect water quality.

Biodiversity Loss:

Species Decline: The combined effects of radiation exposure, habitat destruction, and contamination can lead to declines in species populations. Some species may become endangered or even extinct if they are unable to adapt to the changing environment.

Genetic Mutations: Radiation exposure can cause genetic mutations in plants, animals, and microorganisms. While some mutations may be harmless, others can lead to malformations, reduced fertility, or increased susceptibility to disease, further threatening biodiversity.

Case Studies of Environmental Impact

Chernobyl Disaster (1986):

The Exclusion Zone: The Chernobyl disaster created a 30-kilometer exclusion zone around the nuclear plant, where radiation levels remain dangerously high. This area has become a case study in the long-term environmental impact of nuclear fallout, with ongoing soil, water, and wildlife contamination.

Ecological Recovery: Despite the high radiation levels, some areas within the exclusion zone have seen a surprising degree of ecological recovery. The absence of human activity has allowed some species to thrive, although they are living in a highly radioactive environment.

Fukushima Daiichi Disaster (2011):

Marine Contamination: The Fukushima disaster led to significant contamination of the Pacific Ocean, with radioactive isotopes detected in marine life far from the accident site. The long-term impact on marine ecosystems remains a concern, particularly for species that migrate through contaminated waters.

Agricultural Impact: The disaster also had a severe impact on agriculture in the surrounding regions, with farmland rendered unusable and crops destroyed due to contamination. The Japanese government continues to monitor and manage the affected areas to mitigate long-term environmental damage.

Nuclear Testing Sites:

Nevada Test Site (United States): The Nevada Test Site, where numerous nuclear tests were conducted during the Cold War, remains contaminated with radioactive isotopes. The site is an ongoing subject of study, providing insights into the long-term environmental impact of nuclear testing.

Bikini Atoll (Marshall Islands): Bikini Atoll, the site of multiple nuclear tests by the United States, remains highly contaminated. The local population was relocated, and the atoll is largely uninhabitable due to the persistent presence of radioactive isotopes in the soil and water.

Mitigating the Environmental Impact of Nuclear Fallout

Decontamination Efforts:

Soil Remediation: Various techniques, such as soil removal, washing, and the use of chemical agents, can reduce soil contamination. Phytoremediation, using plants to absorb radioactive isotopes, is another promising approach.

Water Treatment: Contaminated water can be treated using filtration systems, reverse osmosis, and chemical treatments to remove radioactive particles. These methods are critical for ensuring safe drinking water and protecting aquatic ecosystems.

Wildlife Protection:

Conservation Programs: Conservation efforts in contaminated areas focus on protecting vulnerable species and monitoring the health of wildlife populations. This includes creating protected areas and conducting research on the long-term effects of radiation on different species.

Radiation Monitoring: Continuous monitoring of radiation levels in the environment helps identify areas of concern and track the spread of contamination. This information is crucial for guiding conservation efforts and protecting both human and animal populations.

Long-Term Environmental Monitoring:

Surveillance Programs: Governments and international organizations conduct long-term surveillance of areas affected by nuclear fallout. This monitoring includes tracking radiation levels in soil, water, and air, as well as studying the health of ecosystems and human populations.

Data Sharing and Collaboration: International collaboration and data sharing are essential for understanding the global impact of nuclear fallout. Organizations such as the International Atomic Energy Agency (IAEA) and the United Nations Environment Programme (UNEP) play key roles in coordinating these efforts.

Preparing for Future Environmental Risks

Climate Change and Nuclear Risks:

Rising Sea Levels: Climate change poses new risks to nuclear facilities, particularly those located near coastlines. Rising sea levels and increased storm activity could lead to more nuclear accidents, exacerbating the environmental impact of nuclear fallout.

Extreme Weather Events: The increasing frequency of extreme weather events, such as hurricanes and wildfires, could compromise the safety of nuclear power plants and storage facilities, leading to potential releases of radioactive material.

Strengthening Environmental Protections:

Regulatory Frameworks: Strengthening international and national regulatory frameworks is essential for minimizing the risk of nuclear fallout and its environmental impact. This includes stricter safety standards for nuclear facilities and better preparedness for nuclear emergencies.

Public Awareness and Education: Educating the public about the environmental risks of nuclear fallout and how to mitigate them is critical for fostering a culture of preparedness and resilience. Public awareness campaigns can help communities understand the long-term consequences of nuclear fallout and the importance of environmental protection measures.

The Role of International Cooperation in Addressing Fallout Impact

Global Monitoring Systems:

International Atomic Energy Agency (IAEA): The IAEA plays a crucial role in monitoring radiation levels globally and providing technical support to countries affected by nuclear fallout. The IAEA also facilitates the sharing of best practices for environmental remediation and disaster response.

Comprehensive Nuclear-Test-Ban Treaty Organization (CTBTO): The CTBTO operates a global network of monitoring stations that detect nuclear explosions and track the spread of radioactive materials. This data is essential for understanding the environmental impact of nuclear fallout and coordinating international response efforts.

Environmental Treaties and Agreements:

Nuclear Non-Proliferation Treaty (NPT): The NPT aims to prevent the spread of nuclear weapons and promote disarmament. By reducing the number of nuclear weapons and testing activities, the treaty indirectly helps to minimize the risk of nuclear fallout and its environmental impact.

Paris Agreement and Climate-Nuclear Nexus: While primarily focused on climate change, the Paris Agreement encourages nations to consider the environmental risks posed by nuclear facilities in the context of a changing climate. Integrating nuclear safety into climate adaptation plans is increasingly recognized as a vital component of global environmental security.

Collaborative Research and Innovation:

Joint Research Initiatives: Collaborative research initiatives between countries, academic institutions, and environmental organizations are essential for developing new technologies and strategies to mitigate the impact of nuclear fallout on the environment. This includes advancements in decontamination, radiation-resistant crops, and environmental monitoring tools.

International Conferences and Workshops: Regular international conferences and workshops bring together experts, policymakers, and stakeholders to discuss the latest developments in nuclear safety and environmental protection. These gatherings are key opportunities for sharing knowledge, coordinating efforts, and building global consensus on best practices.

Preparing for the Long-Term: Sustainable Practices and Policy Development

Sustainable Land Use and Recovery:

Rehabilitation of Contaminated Land: Long-term efforts to rehabilitate land affected by nuclear fallout involve a combination of decontamination, reforestation, and sustainable land management practices. These efforts aim to restore ecosystems and make land safe for future use, whether for agriculture, habitation, or conservation.

Promoting Biodiversity: In areas where radiation levels have diminished, reintroducing native species and promoting biodiversity can help restore ecological balance. Conservation programs that focus on resilient species and ecosystems are critical for the long-term recovery of contaminated environments.

Policy and Regulatory Development:

Strengthening Nuclear Safety Regulations: Policymakers must continually update and strengthen nuclear safety regulations to address emerging threats, such as climate change and technological advancements. This includes stricter oversight of nuclear facilities, enhanced safety protocols, and robust emergency response plans.

Incorporating Environmental Impact Assessments: Environmental impact assessments (EIAs) should be a mandatory component of any nuclear-related project, including the construction of new power plants, waste storage facilities, or military installations. EIAs help identify potential environmental risks and guide the implementation of mitigation measures.

Community Engagement and Resilience:

Empowering Local Communities: Local communities should be actively involved in decision-making processes related to nuclear safety and environmental protection. Empowering communities with knowledge and resources enables them to advocate for sustainable practices and prepare for potential nuclear incidents.

Resilience Planning: Communities at risk of nuclear fallout should develop comprehensive resilience plans that include strategies for protecting the environment, maintaining food and water security, and ensuring public health. These plans should be integrated into broader disaster preparedness and climate adaptation efforts.

The impact of nuclear fallout on the environment is profound and long-lasting, affecting air, soil, water, and living organisms in ways that can persist for decades or even centuries. International cooperation, sustainable practices, and robust policy development are essential for addressing the challenges posed by nuclear fallout. As we move forward, integrating environmental considerations into nuclear safety and disaster preparedness efforts will be critical for protecting both ecosystems and human health in the face of nuclear threats.

Agriculture and Livestock: Protecting Your Resources

Agriculture and livestock are vital components of survival during and after a nuclear fallout event. Protecting these resources is essential for ensuring food security and maintaining long-term self-sufficiency in a fallout shelter scenario. This chapter explores strategies for safeguarding crops and livestock from radioactive contamination, managing resources effectively, and ensuring the safety and sustainability of food supplies in a post-fallout environment.

Understanding the Risks to Agriculture and Livestock

Radioactive Contamination:

Direct Fallout Exposure: Crops and livestock can be directly exposed to radioactive fallout as particles settle on the ground, plants, and animals. This contamination poses significant health risks and can render food supplies unsafe for consumption.

Soil and Water Contamination: Radioactive particles can seep into the soil, contaminating crops and groundwater supplies. Contaminated soil can take decades to recover, affecting crop yields and the safety of agricultural products for years.

Bioaccumulation: Radioactive isotopes can accumulate in plants and animals over time. This process, known as bioaccumulation, results in higher concentrations of radiation in the food chain, posing greater risks to human health when contaminated food is consumed.

Impact on Livestock:

Health Risks: Livestock exposed to radiation may suffer from acute radiation sickness, reduced fertility, birth defects, and increased mortality. These health issues can drastically reduce the productivity of livestock herds and compromise food security.

Contaminated Feed and Water: Livestock that consume contaminated feed or water are at risk of internal radiation exposure, which can lead to the contamination of meat, milk, eggs, and other animal products.

Protecting Crops from Nuclear Fallout

Pre-Fallout Preparations:

Greenhouse and Indoor Farming: One of the most effective ways to protect crops from fallout is by growing them in controlled environments such as greenhouses or indoor farms. These structures provide a physical barrier against radioactive particles and can be equipped with air filtration systems to maintain a clean environment.

Soil Testing and Amendment: Regularly test soil for radiation levels and amend it with clean, uncontaminated soil as needed. Using raised beds with imported soil can also help protect crops by reducing their exposure to contaminated ground.

Cover Crops and Mulching: Plant cover crops and apply mulch to protect the soil from direct fallout contamination. These practices help shield the soil, reduce erosion, and maintain soil health.

Post-Fallout Actions:

Decontamination of Plants: If crops are exposed to fallout, they can be washed with water to remove some of the radioactive particles from their surfaces. However, this method is only partially effective and may not completely eliminate the contamination.

Selective Harvesting: In cases where only part of the crop is contaminated, it may be possible to selectively harvest unaffected areas. However, thorough testing is essential to ensure that the harvested produce is safe for consumption.

Crop Rotation and Fallow Periods: Rotate crops and allow fields to lie fallow for a period to reduce the buildup of radioactive isotopes in the soil. Over time, some radioactive isotopes will decay and reduce their harmful impact, making the land safer for future cultivation.

Radiation-Resistant Crops:

Choosing Hardy Varieties: Some crop varieties are more resistant to environmental stressors, including radiation. Research and select crops that are known for their hardiness, such as certain types of grains, legumes, and root vegetables.

Breeding and Genetic Engineering: Advances in breeding and genetic engineering may produce crops that are more resistant to radiation. While this is an emerging field, it holds potential for improving agricultural resilience in contaminated environments.

Safeguarding Livestock

Sheltering Livestock:

Indoor Housing: Keeping livestock indoors during and after a fallout event is crucial for minimizing exposure to radioactive particles. Sturdy barns or underground shelters equipped with ventilation and air filtration systems can provide a safe environment for animals.

Feed and Water Storage: Store a supply of clean feed and water in sealed containers to ensure that livestock have access to uncontaminated food and water. This supply should be sufficient to last for several weeks to months, depending on the anticipated duration of fallout effects.

Health Monitoring and Veterinary Care:

Radiation Sickness Symptoms: Monitor livestock for signs of radiation sickness, such as lethargy, loss of appetite, skin lesions, and abnormal behavior. Early detection and intervention are key to managing health issues and preventing the spread of contamination.

Veterinary Support: Ensure that you have access to veterinary care or have basic veterinary knowledge to treat common ailments and manage the health of your animals. Stockpile veterinary supplies, including medications, bandages, and antiseptics.

Breeding and Genetic Management:

Selective Breeding: After a fallout event, it may be necessary to selectively breed livestock to ensure the continuation of healthy, radiation-free animals. Focus on breeding animals that have shown resilience to environmental stressors and maintain genetic diversity to prevent inbreeding.

Culling Contaminated Animals: In cases of severe contamination, it may be necessary to cull animals that have been heavily exposed to radiation to prevent the spread of contamination through the food chain.

Managing Food Resources from Agriculture and Livestock

Testing for Contamination:

Radiation Detectors: Use radiation detectors to test crops, animal products, and soil for contamination. Regular testing is essential for ensuring that food is safe for consumption and for monitoring radiation levels over time.

Sample Testing: Collect and test samples from different areas of your farm or garden to identify hotspots of contamination. This information can guide decisions about which areas to harvest, abandon, or decontaminate.

Food Preservation Techniques:

Canning and Pickling: Preserve safe, uncontaminated crops and animal products through canning and pickling. These methods extend the shelf life of food and provide a buffer against future food shortages.

Drying and Smoking: Drying and smoking meats and produce are effective ways to preserve food without refrigeration. These methods can also help reduce the volume of food, making it easier to store and transport.

Rationing and Resource Management:

Rationing Plans: Develop a rationing plan that ensures fair distribution of food resources among shelter occupants. Rationing helps extend food supplies and reduces waste during periods of scarcity.

Rotating Stock: Regularly rotate stored food supplies to ensure freshness and prevent spoilage. Implement a first-in, first-out (FIFO) system to use older supplies before newer ones.

Long-Term Strategies for Agricultural and Livestock Recovery

Soil Remediation:

Phytoremediation: Use plants that absorb radioactive isotopes to clean contaminated soil. These plants can be harvested and safely disposed of, reducing the radiation levels in the soil over time.

Soil Amendment and Replacement: Add clean soil or amendments like lime, clay, or organic matter to contaminated fields to reduce radiation levels and improve soil health. In severe cases, replacing the topsoil may be necessary.

Sustainable Agriculture Practices:

Crop Diversification: Diversify crops to reduce the risk of total crop failure and to improve soil health. A varied crop rotation plan helps maintain nutrient balance and reduces the impact of any one crop being contaminated.

Regenerative Farming: Implement regenerative farming practices, such as no-till farming, cover cropping, and composting, to restore soil health and increase resilience to environmental stressors.

Livestock Recovery Programs:

Restocking Programs: If livestock populations are severely impacted, consider participating in restocking programs that provide healthy animals to replenish herds. These programs often focus on restoring genetic diversity and rebuilding sustainable livestock populations.

Animal Husbandry Education: Invest in education and training on modern animal husbandry techniques that prioritize health, productivity, and resilience. This knowledge can help improve livestock management in a post-fallout environment.

Planning for Future Agricultural Resilience

Emergency Preparedness Plans:

Agricultural Contingency Planning: Develop comprehensive contingency plans that outline how to protect, manage, and restore agricultural and livestock resources in the event of a nuclear fallout. These plans should include protocols for sheltering animals, testing for contamination, and decontaminating affected areas.

Community Collaboration: Work with neighboring farms and communities to develop shared resources and strategies for protecting agriculture and livestock. Collaborative efforts can enhance resilience and provide mutual support during and after a crisis.

Research and Innovation:

Investing in Agricultural Research: Support research into radiation-resistant crops, livestock breeds, and soil remediation techniques. Innovation in these areas can significantly improve the resilience of agriculture and livestock in the face of nuclear threats.

Adopting New Technologies: Embrace new technologies that enhance agricultural resilience, such as precision farming tools, automated monitoring systems, and genetically engineered crops. These technologies can help farmers better manage resources and respond to environmental challenges.

Protecting agriculture and livestock from nuclear fallout is essential for ensuring food security and long-term sustainability in a post-fallout environment. By understanding the risks, implementing protective measures, and planning for recovery, individuals and communities can safeguard their vital resources and maintain a stable food supply. Effective management of crops and livestock, combined with ongoing monitoring and testing, will be key to navigating the challenges posed by radioactive contamination. Through careful planning, sustainable practices, and collaboration, it is possible to build a resilient agricultural system that can withstand and recover from the impacts of nuclear fallout.

Prepping for a Nuclear War: What to Expect

Prepping for a nuclear war involves understanding the potential scenarios, the immediate and long-term consequences, and the steps necessary to protect yourself and your family. This chapter will guide you through what to expect before, during, and after a nuclear event, including practical tips for preparing your home, building a fallout shelter, stocking essential supplies, and maintaining mental and physical well-being during such a crisis.

Understanding the Threat: Nuclear War Scenarios

Types of Nuclear Events:

Full-Scale Nuclear War: A full-scale nuclear war involves the widespread use of nuclear weapons by two or more nations. This scenario could result in multiple detonations across various regions, leading to catastrophic loss of life, infrastructure, and environmental damage. The fallout from such a war could affect the entire planet.

Limited Nuclear Exchange: A limited nuclear exchange refers to the use of a small number of nuclear weapons in a localized conflict. While less devastating than a full-scale war, the consequences would still be severe, with significant regional fallout, destruction, and long-term environmental impact.

Nuclear Terrorism: The detonation of a nuclear device by a non-state actor, such as a terrorist group, poses a different kind of threat. This scenario might involve a single nuclear explosion in an urban area, resulting in localized but intense destruction and fallout.

Warning Systems and Detection:

Government Alert Systems: Governments around the world have established warning systems to detect and alert the public to an impending nuclear attack. These systems might include sirens, emergency broadcast messages, and mobile alerts. Familiarize yourself with the alert systems in your area and ensure that you are registered for any available notification services.

Time to React: In many cases, the time between a warning and the detonation of a nuclear weapon could be as short as 10-30 minutes. Immediate action is required, with no time to gather supplies or make last-minute preparations.

Preparing Your Home and Shelter

Building or Selecting a Fallout Shelter:

Types of Shelters: The safest option is to have a dedicated fallout shelter, either underground or heavily reinforced, to protect against the blast and radiation. If constructing a new shelter is not feasible, designate a basement or interior room as your shelter space, reinforcing it as much as possible with heavy materials like concrete, lead, or earth.

Stocking Your Shelter: Your shelter should be stocked with enough supplies to last at least two weeks, though longer is preferable. Essential supplies include non-perishable food, water, medical supplies, radiation detectors, and sanitation items. Include items for comfort and mental health, such as books, games, or other entertainment.

Ventilation and Air Filtration: Ensure that your shelter has proper ventilation and an air filtration system to remove radioactive particles. If your shelter is underground, consider installing an air pump or manual ventilation system to maintain air quality.

Home Preparations:

Sealing Your Home: If you do not have access to a dedicated fallout shelter, you can prepare your home by sealing windows, doors, and vents with plastic sheeting and duct tape. This helps to prevent radioactive dust from entering your living space.

Water Storage: Store large quantities of clean water, as contamination of water supplies is likely in the aftermath of a nuclear event. Consider storing water in food-grade containers and learning how to purify water using filtration, boiling, or chemical treatments.

Emergency Power: In the event of a nuclear war, power grids are likely to fail. Prepare for this by having backup power sources, such as a generator, solar panels, or batteries. Ensure that you have enough fuel and supplies to keep essential systems running.

Stockpiling Essential Supplies

Food and Water:

Non-Perishable Food: Stock up on non-perishable foods that are easy to store and prepare, such as canned goods, dried fruits, nuts, rice, pasta, and freeze-dried meals. Ensure that you have a variety of foods to meet nutritional needs over an extended period.

Water Supply: Plan for at least one gallon of water per person per day. Include additional water for sanitation, cooking, and pets. Consider using water storage barrels or other large containers that can be easily accessed and refilled.

Food Preservation: Learn how to preserve food through methods like canning, drying, and smoking. This knowledge will be invaluable in extending your food supplies over the long term.

Medical and First Aid Supplies:

First Aid Kit: A comprehensive first aid kit should include bandages, antiseptics, pain relievers, burn ointments, splints, and other basic medical supplies. Consider including additional items for radiation exposure, such as potassium iodide tablets, which can help protect the thyroid from radioactive iodine.

Prescription Medications: Ensure that you have an adequate supply of any prescription medications needed by family members. Work with your healthcare provider to obtain a longer-term supply if possible.

Radiation Detection: Equip your shelter with radiation detection devices, such as Geiger counters or dosimeters, to monitor radiation levels and determine when it is safe to leave the shelter.

Sanitation and Hygiene:

Portable Toilets: If your shelter does not have a built-in toilet, consider a portable toilet or bucket system with plastic liners and absorbent materials. Store enough supplies to manage waste for the duration of your stay in the shelter.

Hygiene Supplies: Stock up on personal hygiene items, including soap, hand sanitizer, toothpaste, toilet paper, and feminine hygiene products. These items are essential for maintaining health and morale during an extended stay in a shelter.

Waste Disposal: Plan for waste disposal by having garbage bags, disinfectants, and methods for safely storing and disposing of waste. Keeping your living space clean and free from waste is crucial for preventing disease.

Mental and Physical Health During a Nuclear Event

Managing Stress and Anxiety:

Mental Health Preparation: The psychological impact of a nuclear event can be severe. Prepare by learning stress management techniques, such as deep breathing, meditation, and mindfulness. Maintaining a sense of routine and engaging in hobbies or entertainment can also help alleviate stress.

Social Support: If possible, maintain communication with loved ones and neighbors. Social support is critical for mental well-being, especially during prolonged confinement. Use radios or other communication devices to stay connected with the outside world.

Physical Fitness:

Exercise in Confined Spaces: Limited space in a shelter can make it challenging to stay physically active, but regular exercise is essential for maintaining health and morale. Engage in bodyweight exercises, stretching, or yoga to keep your body moving.

Nutrition and Hydration: Proper nutrition and hydration are crucial for maintaining physical health. Ensure that you are consuming balanced meals and drinking enough water to stay hydrated, even in stressful conditions.

Dealing with Illness and Injury:

First Aid Skills: Basic first aid skills are essential for treating minor injuries and managing medical emergencies during a nuclear event. Consider taking a first aid course and practicing key skills, such as CPR, wound care, and splinting.

Quarantine and Infection Control: In the event of illness, isolate the affected individual as much as possible to prevent the spread of infection. Use protective gear, such as masks and gloves, when caring for sick family members.

What to Expect During and After a Nuclear Event

The Immediate Aftermath:

Blast and Heat Effects: The immediate effects of a nuclear explosion include a blinding flash of light, intense heat, and a shockwave capable of destroying buildings and infrastructure. If you are within range, seek shelter immediately and avoid looking at the blast.

Fallout and Radiation: Radioactive fallout will begin to settle within minutes to hours after the explosion. It is critical to stay indoors, preferably in a well-protected shelter, to avoid exposure to this fallout. Radiation levels will be highest immediately following the explosion and will decrease over time, though some areas may remain hazardous for years.

Surviving the First Few Days:

Shelter in Place: Stay in your shelter for at least 48 hours, as this is the period when radiation levels are most dangerous. After this time, use a radiation detector to assess whether it is safe to leave the shelter for short periods.

Communication and Information: Tune into emergency broadcasts or use a battery-powered radio to stay informed about the situation outside. Government agencies may provide updates on radiation levels, safe zones, and relief efforts.

Long-Term Considerations:

Leaving the Shelter: Once radiation levels have decreased to safe levels, you may be able to leave the shelter for limited periods. However, caution is still necessary, as some areas may remain hazardous. Continue to monitor radiation levels and follow government guidance.

Rebuilding and Recovery: The aftermath of a nuclear event will involve significant challenges, including rebuilding homes, restoring infrastructure, and dealing with long-term environmental contamination. Be prepared for a slow and difficult recovery process, and consider working with your community to share resources and support each other.

Building a Long-Term Survival Plan

Sustainable Living:

Food Production: Consider long-term strategies for producing your own food, such as starting a garden (once the soil is safe) or raising small livestock. These activities can help sustain you and your family during prolonged periods of instability.

Water Security: Develop a plan for securing a reliable source of clean water, such as installing a well, harvesting rainwater, or purifying water from natural sources. Water security will be one of the most critical aspects of long-term survival.

Skills Development:

Self-Reliance Skills: Invest time in learning self-reliance skills, such as gardening, hunting, fishing, and basic construction. These skills will be invaluable in a post-nuclear world where access to goods and services may be limited. Additionally, learning how to repair and maintain your shelter, tools, and equipment can help you stay self-sufficient.

First Aid and Medical Training: Expanding your knowledge of first aid and basic medical care is crucial, especially in a situation where professional medical help might be unavailable. Consider taking advanced first aid courses, learning how to use medical equipment, and understanding how to treat common ailments and injuries that could arise during long-term survival.

Bartering and Trade:

Valuable Skills and Items: In a post-nuclear world, traditional currency may lose its value, and bartering could become a primary means of obtaining goods and services. Skills such as carpentry, mechanics, and medicine, as well as items like clean water, food, and tools, could become valuable commodities for trade.

Building Community Networks: Establish connections with neighbors and other survivors to form a network of mutual support and trade. Communities that work together to pool resources and share skills will be better equipped to navigate the challenges of a post-nuclear landscape.

Mental and Emotional Resilience:

Coping with Trauma: The psychological toll of surviving a nuclear event can be immense, with survivors potentially facing trauma, grief, and uncertainty. Building mental resilience is just as important as physical preparation. Engage in practices like journaling, meditation, and staying connected with loved ones to manage stress and emotions.

Maintaining Hope and Purpose: In the aftermath of a nuclear war, finding hope and purpose is vital for long-term survival. Setting goals, maintaining routines, and focusing on the well-being of your family and community can provide the motivation needed to persevere through difficult times.

Preparing for Different Scenarios

Urban vs. Rural Survival:

Urban Challenges: In urban areas, the density of buildings and population can make surviving a nuclear event more challenging due to the higher likelihood of fallout, limited access to clean water, and potential for civil unrest. Urban preppers should focus on building strong, well-stocked shelters and having plans for evacuation if necessary.

Rural Advantages: Rural areas may offer more opportunities for self-sufficiency, such as access to land for farming and fewer immediate threats from fallout. However, rural preppers need to ensure that they have the resources and skills to survive in relative isolation.

Nuclear Winter Considerations:

Extended Shelter Time: A nuclear winter, caused by the soot and debris from nuclear explosions blocking sunlight, could lead to drastically lower temperatures and reduced food production worldwide. In such a scenario, you may need to stay in your shelter for an extended period, possibly months or longer.

Cold Weather Survival: Prepare for cold weather by ensuring your shelter is well-insulated, and stockpile winter clothing, blankets, and heating sources. Learning how to start fires, create makeshift stoves, and maintain warmth in your shelter will be critical for survival during a nuclear winter.

Final Steps: Reviewing and Refining Your Plan

Regular Drills and Practice:

Shelter Drills: Regularly practice sheltering-in-place drills with your family to ensure that everyone knows what to do in the event of a nuclear attack. These drills should include sealing the shelter, managing supplies, and using radiation detection equipment.

Scenario-Based Planning: Consider different nuclear war scenarios and tailor your preparations accordingly. Practice what to do if you are at work, school, or away from home when a nuclear event occurs, and ensure that your family has a plan for reunification.

Continuous Learning:

Staying Informed: Stay up-to-date on global events, nuclear policies, and advancements in nuclear safety. Understanding the geopolitical landscape and potential risks can help you anticipate threats and adjust your preparations as needed.

Knowledge Refreshment: Regularly refresh your knowledge of nuclear survival techniques, first aid, and other essential skills. Staying sharp and informed will ensure that you can respond effectively in an emergency.

Adapting to Changing Circumstances:

Flexible Planning: Be prepared to adapt your plans as circumstances change. This could involve moving to a new location, adjusting your supply stockpile, or rethinking your shelter design based on new information or experiences.

Community Preparedness: Work with your community to develop a collective preparedness plan. Having a network of trusted individuals can provide additional security, resources, and support in the event of a nuclear war.

Prepping for a nuclear war is a complex and challenging task that requires careful planning, resource management, and mental resilience. By understanding the potential scenarios, preparing your home and shelter, stockpiling essential supplies, and maintaining your mental and physical health, you can increase your chances of surviving and thriving in a post-nuclear world.

Remember, survival is not just about enduring the immediate aftermath of a nuclear event; it's also about planning for the long term, rebuilding your life, and helping your community recover. With thorough preparation and a resilient mindset, you can navigate the challenges of a nuclear war and protect yourself and your loved ones from the worst effects of such a catastrophic event.

Long-Term Survival: Beyond the Initial Fallout

Long-term survival after a nuclear event goes beyond the initial fallout and requires sustained efforts to adapt, rebuild, and thrive in a drastically changed environment. Once the immediate danger of radiation exposure has passed, the focus shifts to maintaining physical and mental health, securing food and water, rebuilding shelter and infrastructure, and fostering community resilience. This chapter will explore strategies for long-term survival, addressing the challenges that arise after the initial fallout and providing practical guidance for navigating life in a post-nuclear world.

Assessing the Environment Post-Fallout

Radiation Levels and Safe Areas:

Ongoing Radiation Monitoring: Even after the initial fallout has settled, radiation levels can remain dangerously high in certain areas. Use radiation detection equipment, such as Geiger counters and dosimeters, to continuously monitor your environment. Understanding the radiation levels in your area will help you make informed decisions about where to live, work, and gather resources.

Mapping Safe Zones: Identify and map out areas with lower radiation levels where it is safer to spend extended periods. These safe zones can be used for farming, gathering resources, and potentially rebuilding homes and communities. Avoid known hotspots where radiation levels are persistently high.

Environmental Changes:

Impact of Nuclear Winter: If a nuclear winter has occurred, you may face dramatically colder temperatures, reduced sunlight, and longer periods of darkness. These conditions can severely impact agriculture, water supplies, and overall living conditions. Prepare to adapt your survival strategies to cope with these extreme changes.

Soil and Water Contamination: Assess the extent of soil and water contamination in your area. Contaminated soil may require remediation before it is safe for growing crops, and water sources may need to be filtered or treated before consumption. Regular testing and purification of water are critical for long-term survival.

Sustaining Food and Water Supplies

Long-Term Food Production:

Agriculture: Once radiation levels allow, start cultivating crops that are resilient and capable of growing in potentially contaminated soil. Root vegetables, hardy grains, and legumes are good options, as they are relatively easy to grow and provide essential nutrients. Implement crop rotation, soil amendments, and phytoremediation techniques to gradually improve soil health.

Livestock: If possible, raise small livestock such as chickens, rabbits, or goats. These animals can provide a steady supply of protein through meat, eggs, and milk. Ensure that their feed and water sources are free from contamination, and monitor their health regularly.

Foraging and Hunting: In some areas, foraging for wild plants and hunting may be viable options for supplementing your food supply. Be cautious, as plants and animals in contaminated areas can carry radioactive isotopes. Use radiation detection equipment to assess the safety of foraged or hunted food.

Water Security:

Water Collection Systems: Install or maintain systems for collecting and storing water, such as rainwater harvesting systems, wells, and cisterns. Ensure that all collected water is filtered and treated to remove radioactive particles and other contaminants.

Purification Techniques: Utilize multiple water purification methods, including filtration, boiling, and chemical treatments, to ensure that your water is safe to drink. Regularly test your water sources for radiation and other pollutants to maintain a reliable and clean water supply.

Food Preservation and Storage:

Canning and Pickling: Preserve food through canning and pickling to extend its shelf life and ensure that you have a steady supply of food during times of scarcity. Properly sealed containers can keep food safe from contamination for months or even years.

Drying and Smoking: Drying and smoking meats, fruits, and vegetables are effective ways to preserve food without the need for refrigeration. These methods also reduce the volume of food, making it easier to store and transport.

Grain Storage: Store grains, legumes, and other dry foods in airtight containers to protect them from moisture, pests, and contamination. Grains are an excellent long-term food source due to their high calorie content and long shelf life.

Rebuilding Shelter and Infrastructure

Shelter Maintenance and Improvements:

Repairing Damage: Over time, your shelter may sustain damage from environmental factors, wear and tear, or past radiation exposure. Regularly inspect and repair any damage to ensure that your shelter remains secure and habitable. Pay special attention to structural integrity, insulation, and ventilation systems.

Expanding Shelter Capacity: As conditions improve, consider expanding your shelter to accommodate more people or provide additional space for food storage, water collection, or livestock. Building additional rooms or outbuildings can increase your long-term survival capacity.

Energy and Power Solutions:

Renewable Energy Sources: In a long-term survival scenario, reliance on renewable energy sources such as solar panels, wind turbines, and hydropower becomes increasingly important. These energy sources can provide a sustainable way to power essential systems like lighting, heating, and water purification.

Energy Conservation: Practice energy conservation by using energy-efficient appliances, reducing power usage during the day, and storing excess energy for nighttime or emergency use. Efficient use of energy resources will extend the lifespan of your equipment and reduce the need for frequent repairs or replacements.

Rebuilding Infrastructure:

Community Projects: Work with other survivors to rebuild essential infrastructure, such as roads, communication networks, and utility services. Collective efforts can accelerate recovery and improve living conditions for everyone.

Reusing and Repurposing Materials: In a post-nuclear environment, new materials may be scarce, so focus on reusing and repurposing existing materials. Salvage building materials from damaged structures, repurpose old equipment, and recycle resources to reduce waste and maximize their utility.

Maintaining Health and Hygiene

Healthcare and Medical Supplies:

Medical Stockpile Management: Continuously manage and replenish your stockpile of medical supplies, including first aid kits, prescription medications, and over-the-counter remedies. Establish a system for rotating supplies to ensure that they remain effective.

Basic Healthcare Skills: In the absence of professional medical help, basic healthcare skills become vital. Learn how to treat common injuries and illnesses, administer first aid, and perform basic medical procedures. Having someone in your group with advanced medical knowledge, such as a nurse or paramedic, can be invaluable.

Sanitation and Waste Management:

Sanitation Practices: Maintaining cleanliness and sanitation is critical for preventing the spread of disease. Regularly clean your living spaces, dispose of waste properly, and ensure that water sources are protected from contamination.

Waste Disposal: Develop a waste disposal system that minimizes environmental impact and prevents contamination of food and water supplies. Composting organic waste, burying non-hazardous waste, and properly sealing hazardous materials are effective practices.

Mental Health and Well-Being:

Coping with Isolation: Long-term survival can lead to feelings of isolation, depression, and anxiety. Combat these feelings by maintaining social connections, establishing routines, and engaging in activities that provide a sense of purpose and normalcy.

Mental Resilience Training: Practice mental resilience techniques, such as mindfulness, meditation, and positive thinking, to help you cope with stress and uncertainty. Encourage open communication within your group to address emotional challenges and provide mutual support.

Community Building and Social Structures

Forming Alliances:

Building Trust: Establish trust with other survivors by sharing resources, knowledge, and skills. Trust is essential for forming strong alliances and ensuring that everyone works together for the common good.

Barter and Trade: Engage in barter and trade with other groups or individuals to acquire needed supplies or services. Bartering can also strengthen relationships and provide access to resources that may be scarce in your immediate area.

Establishing Social Structures:

Leadership and Governance: As your group grows, consider establishing a leadership structure or council to make decisions, resolve disputes, and coordinate efforts. Clear communication and fair decision-making processes are key to maintaining harmony and cooperation.

Community Norms and Values: Develop a set of community norms and values that guide behavior, resource sharing, and conflict resolution. These norms help create a sense of order and shared purpose, which is vital for long-term survival.

Education and Skill Development:

Knowledge Sharing: Encourage the sharing of knowledge and skills within your community. Teaching others essential survival skills, such as farming, hunting, first aid, and mechanics, helps ensure that everyone contributes to the group's well-being.

Lifelong Learning: Promote a culture of lifelong learning by continuously seeking out new knowledge and skills. Staying adaptable and open to new ideas will help you and your community navigate the challenges of a changing world.

Long-Term Planning and Adaptation

Adapting to Changing Conditions:

Flexibility in Planning: Long-term survival requires the ability to adapt to changing conditions, whether due to environmental factors, resource availability, or social dynamics. Regularly reassess your situation and be prepared to modify your plans as needed.

Contingency Plans: Develop contingency plans for various scenarios, such as relocating to a new area, dealing with resource shortages, or responding to new threats. Having backup plans in place ensures that you are not caught off guard by unexpected developments.

Sustainable Practices:

Environmental Stewardship: Practice environmental stewardship by using resources responsibly, minimizing waste, and protecting natural habitats. Sustainable practices not only ensure the long-term health of your environment but also support the regeneration of resources.

Self-Sufficiency: Aim for self-sufficiency in food, water, energy, and shelter. The more self-sufficient you become, the less reliant you are on external resources, which may be scarce or unavailable in a post-nuclear world.

Looking to the Future:

Rebuilding Civilization: As conditions improve and your community becomes more stable, consider the long-term goal of rebuilding civilization on a larger scale. This may involve developing infrastructure, education systems, governance structures, and cultural institutions that support a thriving society. While the path to rebuilding will be challenging, focusing on these efforts can help restore a sense of normalcy and progress.

Passing on Knowledge:

Education for Future Generations: As your community grows and new generations are born, it's essential to pass on the knowledge and skills that have been crucial for survival. Establish educational programs that teach children about survival, agriculture, science, and history, ensuring that they are prepared to take on the challenges of the future.

Documenting Experiences: Record your experiences, challenges, and solutions in a form that can be preserved and shared with others. This documentation could be in written journals, digital media (if available), or oral traditions. By sharing your knowledge, you contribute to the collective wisdom that future generations can draw upon.

Building a Resilient Community:

Social Cohesion: A strong, cohesive community is better equipped to face challenges together. Promote inclusivity, cooperation, and mutual support within your group. Celebrate successes and milestones, and find ways to maintain morale, even in difficult times.

Cultural Preservation and Innovation: Encourage the preservation of cultural practices, arts, and traditions that give meaning to life and foster a sense of identity. At the same time, be open to innovation and new ways of doing things that can improve your community's resilience and quality of life.

Preparing for Future Generations

Establishing Legacy Goals:

Vision for the Future: As a community, develop a shared vision for the future. This vision can include goals for growth, sustainability, education, and cultural development. A clear, collective purpose will guide your community's efforts and keep everyone focused on positive long-term outcomes.

Intergenerational Planning: Plan not just for immediate survival but for the well-being of future generations. Consider how your actions today will impact the environment, resources, and social structures that your children and grandchildren will inherit.

Resource Management and Regeneration:

Sustainable Resource Use: Implement practices that ensure the sustainable use of resources. For example, practice rotational farming, sustainable fishing, and responsible hunting to avoid depleting natural resources. Reforesting areas and managing water sources responsibly will help regenerate ecosystems.

Renewable Energy and Technology: Invest in renewable energy technologies that can provide a continuous power supply without depleting resources. Solar panels, wind turbines, and bioenergy systems can be the backbone of a self-sufficient energy strategy. Additionally, explore new technologies that can improve living conditions, agriculture, and communication.

Continuing Education and Innovation:

Fostering Innovation: Encourage a culture of innovation where community members are motivated to find new solutions to ongoing challenges. Innovation can come from improving tools, creating new farming techniques, or developing more efficient energy systems.

Educational Institutions: If conditions allow, consider establishing formal educational institutions where structured learning can take place. Education is key to societal development and will be crucial for training future leaders, engineers, doctors, and other essential roles.

Long-term survival after a nuclear event requires more than just getting through the initial fallout; it demands resilience, adaptability, and a forward-thinking mindset. As the immediate threats subside, the focus shifts to rebuilding, not just to survive but to thrive. Your efforts today will shape the world that future generations inherit. Through careful planning, sustainable practices, and a commitment to community, it is possible not only to survive a nuclear catastrophe but to emerge from it stronger, more united, and with a renewed sense of purpose. As you plan for the long-term, remember that the strength of your community, the sustainability of your practices, and the resilience of your spirit are the keys to thriving in a post-nuclear world.

Community Planning: Working Together for Safety

Community planning is essential for ensuring safety, resilience, and long-term survival in the aftermath of a nuclear event. By working together, communities can pool resources, share knowledge, and create systems that support collective well-being. This chapter will explore the importance of community planning, strategies for organizing and coordinating efforts, and the benefits of a united approach to safety and survival.

The Importance of Community in a Post-Nuclear World

Shared Resources and Skills:

Pooling Resources: In the aftermath of a nuclear event, resources such as food, water, medical supplies, and tools may be scarce. By pooling resources, communities can ensure that everyone has access to the essentials needed for survival. This collective approach can prevent shortages and reduce waste.

Skill Sharing: Different community members will have different skills, such as medical knowledge, engineering, farming, or hunting. By sharing these skills, the community can function more effectively, with each person contributing to the group's overall survival.

Enhanced Security:

Collective Defense: Working together allows communities to establish a collective defense strategy. This might include organizing watch shifts, creating barriers, or setting up early warning systems. A united community is better equipped to protect itself from external threats, whether they are from other desperate survivors or from environmental dangers.

Shared Information: Communities that communicate and share information can respond more quickly to emerging threats, such as radiation hotspots, contaminated water sources, or the presence of dangerous groups. Staying informed and acting as a unit enhances overall safety.

Psychological and Emotional Support:

Mental Health Benefits: Surviving a nuclear event is not just about physical survival; it also requires maintaining mental and emotional well-being. Being part of a community provides social support, reducing feelings of isolation, fear, and hopelessness. Regular interaction, shared activities, and mutual encouragement can significantly improve morale.

Shared Purpose: A strong community fosters a sense of shared purpose and responsibility. This can help individuals stay motivated and focused on the collective goal of rebuilding and thriving after a disaster.

Organizing Community Efforts

Forming a Community Council:

Leadership Structure: Establish a leadership structure or community council to guide decision-making, coordinate resources, and manage conflicts. The council should include representatives with diverse skills and perspectives to ensure that all voices are heard.

Decision-Making Process: Develop a clear, fair decision-making process. This could be through consensus, majority vote, or a rotating leadership system. The key is to ensure that decisions are made efficiently and reflect the community's needs and values.

Emergency Planning and Drills:

Creating a Community Emergency Plan: Work together to develop a comprehensive emergency plan that addresses potential scenarios, such as further nuclear incidents, resource shortages, or security threats. The plan should include protocols for communication, evacuation, resource distribution, and medical care.

Regular Drills: Conduct regular emergency drills to ensure that all community members know what to do in various situations. These drills should include shelter-in-place exercises, fire drills, first aid training, and communication tests. Practicing these drills helps to identify weaknesses in the plan and keeps everyone prepared.

Resource Management:

Inventory and Distribution: Keep an inventory of all community resources, including food, water, medical supplies, tools, and equipment. Establish a system for fair distribution, ensuring that resources are allocated based on need and availability. Regularly update the inventory to reflect changes in stock.

Sustainable Use: Promote the sustainable use of resources by implementing rationing systems, recycling, and conservation practices. Encourage community members to share ideas and strategies for extending the life of resources and reducing waste.

Building and Maintaining a Safe Community Environment

Establishing Safe Zones:

Designating Safe Areas: Identify and designate safe zones within the community for living, food storage, medical care, and recreation. These zones should be in areas with low radiation levels and be well-protected from external threats. Clear boundaries and signage can help maintain order and safety.

Decontamination Procedures: Develop and enforce decontamination procedures for anyone or anything entering the community from potentially contaminated areas. This includes washing down clothing, equipment, and vehicles, and using radiation detectors to check for contamination.

Community Health and Sanitation:

Public Health Initiatives: Implement public health initiatives to prevent the spread of disease and maintain overall well-being. This includes regular health checks, vaccinations if available, and education on hygiene practices. Addressing health issues early can prevent larger outbreaks.

Sanitation Infrastructure: Build and maintain sanitation infrastructure, such as latrines, waste disposal systems, and clean water facilities. Proper sanitation is crucial for preventing disease and maintaining a healthy living environment.

Conflict Resolution and Governance:

Developing Rules and Norms: Establish community rules and norms that govern behavior, resource use, and interactions. These rules should be developed with input from all community members and should reflect the community's values and priorities.

Conflict Resolution Mechanisms: Create mechanisms for resolving conflicts peacefully and fairly. This could include a mediation committee, regular community meetings, or a system of appeals. Effective conflict resolution helps maintain harmony and trust within the community.

Long-Term Community Planning and Resilience

Education and Skill Development:

Training Programs: Implement training programs that teach essential survival skills, such as farming, first aid, carpentry, and water purification. These programs help ensure that all community members can contribute to long-term survival and resilience.

Mentorship and Apprenticeship: Encourage mentorship and apprenticeship opportunities, where experienced members of the community teach others. This not only preserves critical knowledge but also strengthens bonds between community members.

Economic and Trade Systems:

Barter and Trade Networks: Establish a barter and trade system within the community and with neighboring communities. A well-organized trade network can provide access to goods and services that are scarce or unavailable locally.

Community Projects: Initiate community projects, such as rebuilding infrastructure, creating community gardens, or developing renewable energy sources. These projects provide work, strengthen the community's self-sufficiency, and foster a sense of pride and accomplishment.

Community Growth and Development:

Planning for Expansion: As your community stabilizes, consider planning for growth and expansion. This could involve welcoming new members, building additional shelters, or expanding agricultural areas. Careful planning is needed to ensure that growth is sustainable and does not strain resources.

Fostering Innovation: Encourage innovation and creativity in problem-solving. As new challenges arise, a community that is open to new ideas and willing to adapt is more likely to thrive. Foster a culture of continuous improvement and learning.

Cultural Preservation and Social Cohesion:

Maintaining Traditions: Preserve cultural traditions, rituals, and celebrations that give the community a sense of identity and continuity. These practices can provide comfort and stability in uncertain times.

Building Social Bonds: Promote activities and events that strengthen social bonds, such as communal meals, group projects, and social gatherings. A strong sense of community and belonging is essential for resilience and collective well-being.

Communication and Information Sharing

Establishing Communication Systems:

Internal Communication: Develop a reliable internal communication system for sharing information quickly and efficiently within the community. This could involve radios, message boards, or regular meetings.

External Communication: Establish ways to communicate with neighboring communities, other survivor groups, and any remaining government or aid organizations. Staying connected with the outside world can provide valuable information and support.

Information Management:

Data Collection and Sharing: Collect and share information on radiation levels, resource availability, weather patterns, and other critical factors that impact survival. Having accurate and up-to-date information allows the community to make informed decisions.

Educational Materials: Create and distribute educational materials that help community members understand the risks they face, the importance of preparedness, and the steps they can take to protect themselves and contribute to the community's well-being.

Preparing for Future Challenges

Adaptation and Flexibility:

Continuous Assessment: Regularly assess the community's strengths, weaknesses, opportunities, and threats (SWOT analysis). This helps identify areas for improvement and prepares the community for future challenges.

Flexible Planning: Develop flexible plans that can adapt to changing conditions. This might include contingency plans for new threats, alternative sources of resources, or backup systems for essential infrastructure.

Disaster Recovery and Risk Management:

Risk Mitigation Strategies: Implement strategies to mitigate risks, such as securing food and water supplies, fortifying shelter structures, and maintaining a state of readiness for potential emergencies. Risk management is an ongoing process that requires vigilance and proactive planning.

Disaster Recovery Planning: Prepare for the possibility of additional disasters, whether they are natural, technological, or human-caused. A well-prepared community can recover more quickly and effectively from setbacks.

Leadership Development:

Succession Planning: Develop a succession plan to ensure that leadership roles can be smoothly transitioned if needed. This ensures that the community remains stable and well-managed, even as circumstances change.

Empowering Future Leaders: Encourage the development of leadership skills among community members. This not only strengthens the current leadership but also ensures that future generations are prepared to take on leadership roles.

Community planning is vital for ensuring safety, resilience, and long-term survival in a post-nuclear world. A well-organized and cohesive community is more than just a survival mechanism; it is a source of strength, resilience, and hope. The ability to respond to new challenges, embrace new ideas, and learn from experience will be crucial to the community's long-term success. Remember that the key to survival in a post-nuclear world is not just individual preparation, but collective action and solidarity.

Evacuation vs. Sheltering in Place: Making the Right Decision

In the aftermath of a nuclear event, one of the most critical decisions you may face is whether to evacuate or shelter in place. This decision can have life-or-death consequences, as each option carries its own risks and benefits depending on the specific circumstances. Understanding the factors involved in making this decision, and being prepared to act swiftly, is essential for maximizing your safety and that of your family and community. This chapter will guide you through the considerations, risks, and strategies associated with both evacuation and sheltering in place, helping you to make an informed choice in a crisis.

Understanding the Scenarios: When to Evacuate and When to Shelter

Evacuation Scenarios:

Imminent Nuclear Blast: If you receive credible information that a nuclear blast is imminent and you are within the likely impact zone, evacuation might be necessary if you can reach a safer location before the detonation. However, the window for safe evacuation is often very short—typically less than 30 minutes—depending on the distance and traffic conditions.

High Radiation Levels: If radiation levels in your area are dangerously high and unlikely to decrease quickly, evacuation to a lower-radiation area may be your best option, provided you can do so safely and have a clear destination.

Environmental Hazards: Other environmental hazards, such as fires, floods, or structural damage to your shelter, may also necessitate evacuation. If your current location is no longer tenable, moving to a safer area becomes essential.

Sheltering in Place Scenarios:

Immediate Fallout Risk: If a nuclear detonation has already occurred and you are not in the immediate blast zone, sheltering in place is usually the safest option. The initial fallout is most dangerous in the first 24-48 hours, and staying indoors during this period minimizes your exposure to radiation.

Limited Evacuation Routes: If evacuation routes are likely to be congested, blocked, or unsafe, sheltering in place is generally preferable. Attempting to evacuate under these conditions could expose you to more danger than staying put.

Uncertain Conditions: If you are unsure about the extent of the damage or the safety of other areas, it is often safer to shelter in place until more information becomes available. Rushing to evacuate without a clear plan or destination can lead to unnecessary risks.

Factors to Consider When Deciding to Evacuate or Shelter

Proximity to the Blast Zone:

Blast Radius: Your proximity to the blast zone is a critical factor in deciding whether to evacuate or shelter in place. Those within a few miles of the detonation site may face severe dangers from the blast, heat, and initial radiation. Evacuation might be necessary if you can move quickly enough to escape these hazards.

Radiation Fallout Zones: Understanding the wind patterns and likely fallout zones can help you determine whether you are at risk from radioactive debris. If you are downwind from the blast site, you may face higher fallout levels, making evacuation a potential option.

Time Available:

Time to Impact: If you have advanced warning of a nuclear event, the time available before impact will largely determine your course of action. With only minutes to spare, sheltering in place is often the only viable option. If you have hours or more, evacuation might be possible, but only if you can do so safely and quickly.

Travel Time: Consider the time it will take to reach a safe location. Factor in potential traffic, road conditions, and the possibility of encountering other evacuees. If the travel time exceeds the safe window for evacuation, it is better to shelter in place.

Communication and Information:

Access to Information: Reliable information is crucial in making the right decision. Monitor official sources, such as government alerts, emergency broadcasts, and radiation detectors, to stay informed about the situation. In the absence of clear information, err on the side of caution and consider sheltering in place.

Communication with Family and Community: Ensure that you have a way to communicate with family members and your community to coordinate your decision. Clear communication helps avoid panic and ensures that everyone understands the plan, whether it's to evacuate or shelter.

Resources and Supplies:

Shelter Preparedness: Assess the condition of your shelter and the availability of supplies. If your shelter is well-stocked with food, water, medical supplies, and radiation protection, you are better equipped to shelter in place for an extended period. If supplies are limited, evacuation may become necessary after the initial fallout period.

Evacuation Kit: If evacuation is the best option, having a pre-prepared evacuation kit (also known as a "go-bag") is crucial. This kit should include essentials like water, non-perishable food, medical supplies, important documents, communication devices, and protective clothing. Being prepared allows for a quicker and more organized evacuation.

How to Shelter in Place Effectively

Choosing the Best Location in Your Home:

Basements and Interior Rooms: The safest places to shelter within your home are basements or interior rooms with no windows. These areas offer the most protection from radiation and potential debris. If possible, reinforce these spaces with additional shielding materials like dense earth, concrete, or heavy furniture.

Sealing the Shelter: Use plastic sheeting and duct tape to seal windows, doors, and vents in your shelter space. This helps to prevent radioactive dust from entering your living area. Ensure that you have a way to ventilate the space without compromising the seal.

Minimizing Radiation Exposure:

Shielding: The more material between you and the outside, the better. Pile sandbags, books, mattresses, or other dense materials around your shelter space to increase the shielding effect. Every bit of added protection reduces the radiation dose you receive.

Time and Distance: Remember the principles of time, distance, and shielding. The less time you spend exposed, the greater the distance from the source, and the more shielding you have, the lower your radiation exposure will be. Stay in your shelter for at least 48 hours, or until you receive reliable information that it is safe to emerge.

Sustaining Yourself While Sheltering:

Food and Water: Ration your food and water supplies to last as long as possible. Each person needs about one gallon of water per day for drinking and basic sanitation. Non-perishable, calorie-dense foods are ideal for long-term sheltering.

Sanitation: Maintain sanitation by using portable toilets, waste bags, or makeshift latrines. Keep your living space as clean as possible to prevent illness and maintain morale. Ensure that waste is securely stored and isolated from living areas.

Communication and Information:

Stay Informed: Continuously monitor emergency broadcasts, government alerts, and any other available information sources. Staying informed about radiation levels, weather conditions, and the broader situation will help you make informed decisions about when it is safe to leave your shelter.

Communicate with Others: If possible, keep in touch with family members, neighbors, or community groups. Sharing information and resources can enhance safety and provide emotional support during the sheltering period.

How to Evacuate Safely

Planning Your Evacuation Route:

Identify Safe Locations: Before a crisis occurs, identify potential safe locations to evacuate to, such as a friend or relative's house in a different region, a designated community shelter, or another pre-planned location. Ensure that these places are outside the expected fallout zone and can accommodate your family.

Multiple Routes: Plan multiple evacuation routes to account for potential road closures, traffic, or hazards. Know alternative paths that avoid major highways, which may become congested. Keep maps handy in case GPS or other digital navigation systems fail.

Evacuating Quickly and Efficiently:

Go-Bag Preparation: Have your go-bag ready and accessible at all times. Include essential items such as food, water, clothing, medical supplies, important documents, and communication devices. A well-prepared go-bag allows you to evacuate quickly without wasting time gathering supplies.

Vehicle Preparation: Keep your vehicle in good working condition, with a full tank of gas, emergency supplies, and a first-aid kit. Ensure that your vehicle can accommodate everyone in your household and has enough space for essential supplies. If using public transportation, know the schedules and routes in advance.

Avoiding Hazards During Evacuation:

Radiation Hotspots: Avoid known radiation hotspots, which are areas with high fallout concentrations. If you must pass through a potentially contaminated area, limit your time there, use protective gear, and keep your windows closed. Radiation levels can be monitored using portable detection devices.

Dealing with Crowds and Panic: In a mass evacuation, you may encounter crowds and panic. Stay calm and stick to your plan. Avoid congested areas if possible, and be prepared to take detours or adjust your route to maintain safety and avoid unnecessary delays.

Arriving at Your Destination:

Decontamination: Upon arriving at your destination, take steps to decontaminate yourself and your belongings. Remove outer clothing and seal it in a plastic bag, wash exposed skin with soap and water, and use radiation detection equipment to check for any residual contamination.

Assessing the New Location: Once you've evacuated to a new location, assess its safety and suitability for long-term sheltering. Ensure that you have access to clean water, food, and medical supplies, and that the area is secure from potential threats.

Adapting to Changing Conditions

Reassessing Your Decision:

Ongoing Monitoring: Conditions can change rapidly in the aftermath of a nuclear event. Continuously reassess your situation based on new information, such as changes in radiation levels, weather patterns, or security threats. What was once a safe decision to shelter in place or evacuate might need to be reconsidered if conditions shift unexpectedly.

Flexibility and Contingency Planning:

Be Ready to Adapt: Stay flexible and ready to switch strategies if necessary. If you are sheltering in place but receive credible information that evacuation is now safer, be prepared to leave quickly. Conversely, if you are evacuating and encounter unexpected dangers or blockages, have a backup plan that allows you to find temporary shelter or reroute.

Contingency Plans: Develop contingency plans for different scenarios. For example, if your primary evacuation route is blocked, know secondary routes or safe locations where you can wait until it's safe to proceed. Likewise, if you must shelter in place for longer than anticipated, have a plan for rationing supplies and managing resources.

Post-Evacuation Challenges:

Reintegration: If you evacuate to a community shelter or another group location, be prepared for the challenges of reintegration into a new community setting. There may be different rules, resource-sharing arrangements, and social dynamics to navigate. Stay cooperative and contribute positively to the group.

Returning Home: Once it is deemed safe to return home, do so cautiously. Radiation levels and environmental hazards should be carefully assessed before re-entering your home. Be prepared for the possibility of damaged infrastructure, contamination, or looting, and take necessary precautions.

Long-Term Considerations:

Rebuilding: Whether you have sheltered in place or evacuated, the long-term focus will eventually shift to rebuilding. This could involve repairing or constructing new shelters, restoring infrastructure, or relocating to a safer area. Plan for the long-term needs of your family and community as you move beyond the immediate crisis.

Community Collaboration: Engage with others who have faced similar decisions about sheltering or evacuating. By sharing experiences and lessons learned, you can build a stronger network of support and collective knowledge that benefits everyone in the post-crisis period.

Psychological and Emotional Well-Being

Managing Stress and Anxiety:

Mental Health Practices: The decision to evacuate or shelter in place can be stressful, with significant anxiety over whether you've made the right choice. Practice stress management techniques, such as deep breathing, mindfulness, and maintaining a routine, to help manage these feelings.

Supporting Others: Provide emotional support to family members or others in your community who may be struggling with the decision-making process. Open communication, shared responsibilities, and mutual reassurance can alleviate stress and foster a sense of solidarity.

Dealing with Uncertainty:

Acceptance of Uncertainty: Understand that in a nuclear event, complete certainty may be impossible. Accepting this uncertainty and focusing on what you can control—such as your preparedness, actions, and attitude—can help reduce anxiety and improve decision-making.

Seeking Information: Stay informed but avoid information overload. Rely on trusted sources, and balance the need for information with the importance of staying calm and focused on immediate priorities.

The decision to evacuate or shelter in place during a nuclear event is one of the most critical choices you may face, with profound implications for your safety and survival. By understanding the factors involved, such as proximity to the blast, available time, resource availability, and the evolving situation, you can make informed decisions that maximize your chances of survival.

Remember that both options—evacuating and sheltering in place—require careful planning, preparation, and adaptability. Whether you decide to stay put or move to a safer location, your success will depend on your ability to assess the situation accurately, act quickly, and respond flexibly to changing conditions. Ultimately, the right decision is the one that keeps you and your loved ones safe, and that decision may need to be revisited as new information becomes available. Stay calm, stay informed, and stay prepared, and you will be better equipped to navigate the challenges of a nuclear event with resilience and confidence.

Reinforcing Your Shelter: Upgrades for Maximum Protection

Reinforcing your shelter is crucial for ensuring maximum protection against the various dangers that may arise during and after a nuclear event. Whether you're sheltering in place in a basement, retrofitting an existing structure, or constructing a dedicated fallout shelter, upgrading your space can significantly increase your safety and comfort. This chapter will guide you through the essential upgrades, materials, and strategies to fortify your shelter against radiation, blast effects, and other potential threats.

Understanding the Threats

Radiation:

Direct Radiation: After a nuclear explosion, direct radiation from the blast can pose an immediate threat to those within a certain radius. While you may be outside the immediate blast zone, residual radiation from fallout is a significant long-term hazard that your shelter must protect against.

Fallout Particles: Radioactive fallout consists of dust and debris that becomes radioactive after being blasted into the atmosphere. As these particles settle, they can contaminate everything they touch, including your shelter's exterior. Effective shielding is essential to minimize exposure.

Blast Effects:

Shockwaves: Even if you're outside the direct blast zone, shockwaves can cause significant structural damage. These waves of high-pressure air can shatter windows, collapse walls, and turn debris into deadly projectiles.

Overpressure: The force exerted by a nuclear blast creates overpressure, which can crush buildings, blow doors off their hinges, and cause severe damage to structures. Reinforcing your shelter against these forces is crucial for survival.

Heat and Fire:

Thermal Radiation: The intense heat generated by a nuclear explosion can ignite fires miles away from the blast site. Your shelter must be resistant to fire, both from the external environment and potential internal sources.

Long-Term Threats:

Prolonged Radiation Exposure: The longer you remain in a contaminated area, the more likely you are to suffer from radiation-related health issues. Your shelter should be designed to sustain long-term habitation with minimal radiation exposure.

Resource Depletion: In a prolonged crisis, your shelter must be equipped to handle resource shortages, including food, water, and energy. Reinforcements should include systems for sustainability and self-sufficiency.

Upgrading Your Shelter's Structure

Wall Reinforcement:

Concrete and Earth Barriers: The most effective way to shield against radiation and blast effects is by thickening your shelter's walls with concrete or compacted earth. A minimum thickness of 12 inches of concrete or 36 inches of compacted earth is recommended to significantly reduce radiation exposure.

Sandbags: Sandbags are a practical and relatively inexpensive way to add protection. They can be stacked around the exterior or interior walls to provide additional shielding. Sandbags filled with dense materials, like gravel or soil, offer better radiation protection.

Lead Shielding: For enhanced radiation protection, consider adding lead sheeting to your walls. Lead is highly effective at blocking gamma rays, but it's expensive and heavy, so it should be used selectively in the most critical areas, such as around sleeping quarters or near air vents.

Roof Reinforcement:

Heavy Roofing Materials: The roof of your shelter should be as strong as the walls. Consider reinforcing it with additional layers of concrete or steel. If building from scratch, a concrete slab roof is ideal, supported by steel beams to withstand overpressure and debris impacts.

Earth Covering: Adding a thick layer of earth or gravel on top of your shelter can improve its resistance to radiation and blast effects. This covering acts as a natural insulator and provides an additional barrier against fallout.

Door and Window Reinforcement:

Blast-Resistant Doors: Replace standard doors with blast-resistant ones made of steel or reinforced wood. These doors should have multiple locking mechanisms and seals to prevent air and radioactive particles from entering the shelter.

Window Sealing: Ideally, a fallout shelter should have no windows, as they are weak points in the structure. If windows are unavoidable, reinforce them with steel shutters, sandbags, or thick plexiglass. Ensure that they are tightly sealed to prevent radiation from penetrating.

Ventilation Systems:

Filtered Ventilation: Install a filtered ventilation system that can provide fresh air while blocking radioactive particles. High-efficiency particulate air (HEPA) filters or activated carbon filters are essential for removing contaminants from the air.

Manual Ventilation: In case of power failure, ensure that you have a manual ventilation system, such as hand-crank air pumps or gravity-driven air vents. These systems should be easy to operate and maintain during an extended crisis.

Enhancing Radiation Shielding

Interior Shielding:

Shielding Rooms: Designate specific rooms within your shelter for maximum radiation protection. These should be areas where you spend the most time, such as sleeping quarters or the kitchen. Reinforce these rooms with additional layers of dense material, like lead panels or concrete.

Mobile Shielding: Consider using mobile shielding solutions, like lead blankets or portable barriers, that can be moved as needed to protect against unexpected radiation sources. These are particularly useful in makeshift shelters or if you need to relocate within your shelter.

Floor and Foundation Shielding:

Underground Shelters: If possible, construct your shelter underground. The earth provides excellent natural radiation shielding and protects against blast effects. If your shelter is above ground, reinforce the floor with concrete or additional earth layers.

Raised Flooring: In some cases, creating a raised floor with additional insulation can help reduce radiation exposure from fallout that has settled on the ground. This technique also helps with moisture control, which is essential for long-term habitation.

Airlock Systems:

Entryway Airlocks: Install an airlock system at the entrance to your shelter. This space between the outside and the main shelter area allows you to decontaminate before entering, reducing the risk of bringing radioactive particles inside. Equip the airlock with brushes, water sprays, and storage for contaminated clothing.

Internal Airlocks: If your shelter has multiple rooms, consider installing internal airlocks between high-risk areas (such as an entrance or ventilation shaft) and the living quarters. This additional layer of protection can further minimize contamination risks.

Preparing for Blast and Fire Protection

Blast Wave Mitigation:

Shock Absorption Materials: Use materials that can absorb shockwaves, such as rubber padding, foam insulation, or flexible polymers, in the construction of your shelter. These materials can help reduce the force of a blast wave and protect the structural integrity of your shelter.

Reinforced Entryways: The entryway is a vulnerable point during a blast. Reinforce it with heavy doors and a secure frame, and consider adding a secondary internal door for extra protection. This setup can prevent the blast wave from entering the shelter and causing damage inside.

Fireproofing:

Fire-Resistant Materials: Use fire-resistant building materials such as concrete, brick, and treated wood for your shelter. Avoid flammable materials in the construction and furnishing of your shelter to minimize fire risks.

Fire Suppression Systems: Equip your shelter with fire extinguishers, fire blankets, and a small fire suppression system, such as a water mist system. These tools are essential for quickly addressing any fires that may break out inside the shelter.

Thermal Protection:

Insulation: Proper insulation can protect against the extreme heat generated by a nuclear explosion. Use fire-resistant insulation materials, such as mineral wool or fiberglass, in your walls and ceilings. This insulation also helps maintain a stable temperature inside the shelter, which is crucial for long-term occupancy.

Heat Reflective Coatings: Consider applying heat-reflective coatings to the exterior of your shelter, especially if it's above ground. These coatings can help deflect thermal radiation and reduce the risk of ignition.

Enhancing Long-Term Liveability

Water Supply Systems:

Filtered Water Storage: Store large quantities of water in your shelter, ideally in tanks or barrels made of food-grade materials. Ensure that these storage systems are connected to a filtration system that can remove radioactive contaminants.

Rainwater Harvesting: If possible, integrate a rainwater harvesting system into your shelter. This system should include a filtration and purification process to ensure that the collected water is safe for drinking.

Food Storage and Preparation:

Long-Term Food Storage: Stockpile non-perishable food items that can last for months or years. Store them in a cool, dry, and dark place to maximize their shelf life. Rotate your food supplies regularly to keep them fresh.

Cooking Options: Equip your shelter with a reliable, safe cooking method. Propane stoves, solar cookers, or wood-burning stoves with proper ventilation are good options. Ensure that you have enough fuel stored to last for the duration of your stay in the shelter.

Power and Lighting:

Backup Power Sources: Install backup power systems such as solar panels, wind turbines, or hand-crank generators. These systems should be capable of powering essential devices like communication equipment, ventilation systems, and lighting.

Efficient Lighting: Use energy-efficient LED lights in your shelter to reduce power consumption. Consider battery-operated or solar-powered lights as backup options.

Sanitation and Waste Management:

Portable Toilets: If your shelter lacks a built-in bathroom, consider using a portable toilet system with chemical treatments to manage waste. Ensure that you have enough supplies to maintain sanitation for an extended period.

Waste Disposal: Plan for waste disposal within the shelter by using sealed containers for waste storage. If possible, designate a separate, sealed room for waste to minimize the risk of contamination and odor.

Testing and Maintenance

Regular Inspections:

Structural Integrity: Regularly inspect your shelter for any signs of wear, damage, or structural weaknesses. Check for cracks in the walls or roof, signs of water infiltration, and any deterioration in materials that could compromise the shelter's protection. Schedule these inspections at least twice a year, or more frequently if the shelter is in active use.

Radiation Shielding: Periodically test the effectiveness of your radiation shielding using a Geiger counter or other radiation detection devices. If you notice any increase in radiation levels within the shelter, it may be necessary to add more shielding materials or repair existing barriers.

Ventilation and Air Filtration: Ensure that your ventilation and air filtration systems are functioning correctly. Replace filters as recommended by the manufacturer and check for any blockages or leaks that could reduce the effectiveness of your system.

Maintenance of Systems:

Power Systems: Test your backup power systems regularly to ensure they are operational. This includes running generators, checking battery levels, and ensuring that renewable energy systems like solar panels are charging properly. Keep fuel supplies for generators fresh and replace them as needed.

Water and Sanitation: Inspect your water storage systems for leaks or contamination and ensure that all purification systems are functioning correctly. Test your water regularly to confirm it is safe for drinking. Check the condition of portable toilets and waste disposal systems, and replenish chemical treatments and other supplies as necessary.

Emergency Drills:

Sheltering Drills: Conduct regular drills to ensure that everyone in your household knows how to quickly and safely enter the shelter in an emergency. Practice sealing the shelter, activating ventilation systems, and using emergency equipment. These drills should be done quarterly, at a minimum, to maintain readiness.

Evacuation Protocols: In case your shelter becomes untenable, practice evacuation procedures. Ensure everyone knows the quickest and safest routes to an alternative shelter or safe location. Review go-bags and ensure they are fully stocked and accessible.

Communication Systems:

Testing Communication Devices: Regularly test all communication devices, such as radios, satellite phones, or two-way radios, to ensure they are functioning. Replace batteries as needed and keep a supply of spare batteries or alternative power sources available.

Updating Contact Lists: Maintain an up-to-date list of emergency contacts, including family members, neighbors, and local authorities. Ensure that all communication channels are tested and that everyone knows how to use them in an emergency.

Preparing for Psychological and Emotional Resilience

Creating a Comfortable Living Space:

Interior Comfort: Make your shelter as comfortable as possible to reduce stress during long stays. Include comfortable bedding, seating, and personal items that make the space feel more like home. Consider decorating the shelter with familiar or calming objects, such as photos, artwork, or soft lighting.

Entertainment and Activities: Stock your shelter with entertainment options such as books, board games, cards, or puzzles to help pass the time and maintain morale. If possible, include devices for playing music or watching movies, which can provide a mental escape from the stress of the situation.

Maintaining Mental Health:

Routine and Structure: Establish a daily routine to provide a sense of normalcy and control. Regular activities such as meal times, exercise, and personal hygiene can help maintain a stable mental state.

Mindfulness and Relaxation Techniques: Practice mindfulness, meditation, or relaxation exercises to manage stress and anxiety. Deep breathing exercises, guided meditation, or even simple stretching can help reduce tension and keep your mind focused.

Social Support:

Staying Connected: If you are sheltering with others, ensure regular communication and group activities to maintain social bonds. Open discussions about fears, concerns, and experiences can help alleviate anxiety and build group cohesion.

Community Networks: If possible, establish a communication network with other nearby shelters or communities. Sharing information, resources, and emotional support can be crucial during extended periods of isolation.

Planning for Post-Emergency Scenarios

Exiting the Shelter Safely:

Radiation Monitoring: Before exiting the shelter, use radiation detection equipment to assess the safety of the environment outside. Only leave the shelter once radiation levels have dropped to a safe level, as indicated by reliable sources or your own measurements.

Gradual Re-entry: When it is safe to leave the shelter, do so gradually. Limit initial exposure to the outside environment, and wear protective clothing and masks to reduce the risk of inhaling or coming into contact with residual radioactive particles.

Assessing and Rebuilding:

Damage Assessment: After leaving the shelter, assess the condition of your home and surrounding area. Look for structural damage, contamination, and other hazards. Determine what repairs or rebuilding efforts are necessary and prioritize essential tasks such as restoring power, water, and sanitation.

Long-Term Recovery Planning: Develop a plan for long-term recovery, including rebuilding infrastructure, re-establishing food and water supplies, and reconnecting with the broader community. Consider what resources you will need and how you can obtain them.

Community Collaboration:

Pooling Resources: If possible, work with neighbors or local communities to pool resources and labor for rebuilding efforts. A coordinated approach can help restore normalcy more quickly and efficiently.

Support Networks: Maintain and strengthen the support networks established during the crisis. These relationships can provide ongoing emotional and practical support as you transition from survival mode to long-term recovery.

Reinforcing your shelter for maximum protection involves a comprehensive approach that addresses structural integrity, radiation shielding, blast and fire protection, and long-term liveability. By making strategic upgrades and maintaining your shelter regularly, you can significantly increase your chances of surviving a nuclear event and enduring the long-term challenges that follow. Remember that a well-prepared shelter is not just about physical protection; it also involves planning for psychological resilience and the ability to adapt to changing conditions.

Regular maintenance, emergency drills, and community collaboration are key components of a successful shelter strategy. By investing in these upgrades and practices, you are not only safeguarding your immediate survival but also laying the foundation for a safer, more secure future in the aftermath of a nuclear event.

Dealing with Fallout Anxiety: Mental Health Strategies

Fallout anxiety is a significant concern in the aftermath of a nuclear event, as the fear of radiation exposure, uncertainty about the future, and the stress of long-term sheltering can take a severe toll on mental health. Dealing with this anxiety requires both proactive strategies and ongoing mental health care to maintain emotional well-being during and after the crisis. This chapter will explore mental health strategies for coping with fallout anxiety, offering practical advice on how to manage stress, support others, and build resilience in challenging circumstances.

Understanding Fallout Anxiety

Sources of Anxiety:

Fear of Radiation Exposure: The invisible and pervasive nature of radiation can lead to heightened anxiety, as it's something that cannot be easily detected without specialized equipment. This fear is compounded by the potential long-term health effects of radiation exposure.

Uncertainty and Lack of Control: The unpredictable nature of a nuclear event, including the duration of fallout and its long-term impact on the environment and health, contributes to a profound sense of uncertainty. This lack of control over the situation can exacerbate feelings of anxiety and helplessness.

Isolation and Confinement: Long-term sheltering, especially in confined spaces with limited social interaction, can lead to feelings of isolation, loneliness, and claustrophobia. The stress of confinement can heighten anxiety and lead to a decline in mental health.

Symptoms of Fallout Anxiety:

Physical Symptoms: Anxiety can manifest physically through symptoms such as headaches, muscle tension, fatigue, difficulty sleeping, and gastrointestinal issues. These symptoms can further stress the body, leading to a cycle of worsening anxiety.

Emotional Symptoms: Emotional symptoms of fallout anxiety include persistent worry, fear, irritability, mood swings, and difficulty concentrating. These emotional responses can interfere with daily functioning and overall well-being.

Behavioral Symptoms: Behavioral changes, such as avoiding certain activities, withdrawing from social interactions, or becoming overly vigilant, are common responses to anxiety. These behaviors can lead to further isolation and reinforce anxious feelings.

Immediate Coping Strategies

Breathing and Relaxation Techniques:

Deep Breathing: Practice deep breathing exercises to help calm your nervous system. Focus on taking slow, deep breaths, inhaling through your nose for a count of four, holding for a count of four, and exhaling through your mouth for a count of four. Repeat this cycle several times until you feel more relaxed.

Progressive Muscle Relaxation: Progressive muscle relaxation involves tensing and then slowly releasing each muscle group in your body. Start from your toes and work your way up to your head, paying attention to the sensation of relaxation as you release the tension.

Mindfulness and Grounding:

Mindfulness Meditation: Mindfulness meditation helps you focus on the present moment, reducing anxiety about the future or past. Sit quietly, close your eyes, and focus on your breath, observing each inhalation and exhalation without judgment. If your mind wanders, gently bring it back to your breath.

Grounding Exercises: Grounding exercises can help you stay connected to the present moment. Use the "5-4-3-2-1" technique: identify five things you can see, four things you can touch, three things you can hear, two things you can smell, and one thing you can taste. This technique can help anchor you in the present and reduce anxiety.

Cognitive Behavioral Techniques:

Challenging Negative Thoughts: Cognitive Behavioral Therapy (CBT) techniques can be effective in managing anxiety. When you notice negative or irrational thoughts, challenge them by asking yourself if they are based on facts or assumptions. Reframe these thoughts in a more positive or realistic light.

Positive Affirmations: Use positive affirmations to counteract negative self-talk. Repeating statements like "I am safe in my shelter," "I am prepared and capable," or "I can handle this situation" can help shift your mindset and reduce anxiety.

Building a Supportive Environment

Social Support:

Staying Connected: Maintain regular communication with family members, friends, or other people in your shelter. Social support is a powerful tool for reducing anxiety, as it helps you feel connected and reassured. Share your feelings, concerns, and coping strategies with others, and listen to their experiences as well.

Group Activities: Engage in group activities, such as playing games, cooking meals together, or organizing small projects. These activities provide a distraction from anxiety and foster a sense of community and mutual support.

Creating a Calm Space:

Comfortable Environment: Make your shelter as comfortable and calming as possible. Use soft lighting, warm blankets, and familiar objects like photos or personal mementos to create a sense of safety and comfort. A soothing environment can help reduce anxiety levels.

Routine and Structure: Establish a daily routine that includes regular times for meals, sleep, exercise, and relaxation. A predictable routine can provide a sense of normalcy and control, helping to reduce anxiety and stress.

Mental Health Resources:

Access to Counseling: If possible, access mental health counseling or support services, either in person or through remote means like phone or video calls. Professional counseling can provide additional tools for managing anxiety and coping with the challenges of a post-nuclear environment.

Peer Support Groups: If counseling is not available, consider forming or joining a peer support group within your shelter or community. Sharing experiences and coping strategies with others who are facing similar challenges can be incredibly beneficial.

Long-Term Mental Health Strategies

Exercise and Physical Activity:

Daily Exercise: Regular physical activity is one of the most effective ways to reduce anxiety. Even in a confined space, you can engage in bodyweight exercises, stretching, yoga, or other forms of movement. Exercise releases endorphins, which improve mood and reduce stress.

Group Workouts: Organize group exercise sessions if you are sheltering with others. Group workouts can make exercise more enjoyable and provide social interaction, which is important for mental health.

Diet and Nutrition:

Balanced Diet: Maintain a balanced diet rich in vitamins, minerals, and other nutrients that support mental health. A diet high in whole foods, such as fruits, vegetables, whole grains, and lean proteins, can help stabilize mood and energy levels.

Hydration: Staying hydrated is crucial for mental and physical well-being. Dehydration can exacerbate feelings of anxiety and fatigue, so ensure you drink enough water throughout the day.

Sleep Hygiene:

Consistent Sleep Schedule: Stick to a regular sleep schedule, going to bed and waking up at the same time each day. A consistent sleep routine helps regulate your body's internal clock and improves the quality of your sleep.

Creating a Sleep-Friendly Environment: Make your sleeping area as comfortable and dark as possible. Use earplugs or a white noise machine to block out any disruptive sounds, and avoid stimulating activities, such as watching intense movies or reading distressing news, before bedtime.

Mental Stimulation and Engagement:

Learning New Skills: Engage in mentally stimulating activities, such as learning a new skill or hobby. This could be something practical, like first aid, or something creative, like drawing or writing. Keeping your mind active can help distract from anxiety and build a sense of accomplishment.

Reading and Puzzles: Reading books, solving puzzles, or playing strategy games can keep your mind engaged and provide a healthy escape from worries about the future.

Helping Others Cope

Supporting Family Members:

Open Communication: Encourage open communication with family members, especially children, about their feelings and fears. Reassure them with factual information and remind them that they are safe. Validate their emotions and provide comfort and support.

Role Modeling: Demonstrate healthy coping strategies, such as deep breathing, positive thinking, and maintaining a routine. When others see you handling stress effectively, it can inspire them to adopt similar strategies.

Leadership and Guidance:

Providing Structure: If you are in a leadership role, provide structure and clear guidelines to those under your care. People are more likely to feel secure and less anxious when they know what is expected of them and what to do in different situations.

Encouraging Participation: Involve others in decision-making processes, especially those that impact daily life in the shelter. Feeling involved and having a say in decisions can reduce feelings of helplessness and anxiety.

Addressing Group Anxiety:

Group Discussions: Hold regular group discussions to address any collective concerns and to provide updates on the situation. Transparency and information-sharing can alleviate anxiety by reducing uncertainty.

Conflict Resolution: Tensions and conflicts may arise in a confined setting, exacerbating anxiety. Address conflicts quickly and fairly, using mediation or group problem-solving techniques to resolve issues and restore harmony.

Planning for the Future

Hope and Resilience:

Future Planning: Encourage a focus on the future and the positive steps that can be taken after the immediate crisis is over. Setting long-term goals, even small ones, can provide a sense of purpose and direction, which is vital for mental health.

Resilience Building: Emphasize the development of resilience, both individually and as a group. Discuss past challenges that have been overcome and draw on these experiences to reinforce the belief that you can handle future difficulties as well.

Long-Term Psychological Care:

Post-Traumatic Growth: Understand that experiencing and overcoming a crisis can lead to personal growth. Reflect on what you've learned about yourself and others during the crisis, and how these lessons can be applied to future challenges.

Continued Mental Health Support: Plan for continued mental health care after the immediate crisis has passed. Long-term effects of anxiety and trauma may emerge, and ongoing support is essential for managing these effects. Seek out mental health resources, such as counseling or therapy, once they become available, and stay connected with support networks that can help you navigate the recovery process.

Community Resilience and Healing:

Rebuilding Together: Focus on community healing and resilience as you begin the process of rebuilding after the crisis. Organize community activities, support groups, and collective efforts that foster a sense of unity and shared purpose. This can be a powerful way to help everyone cope with the aftermath and move forward together.

Celebrating Milestones: As your community progresses through recovery, take time to celebrate milestones, no matter how small. Recognizing achievements and progress can boost morale and provide a sense of normalcy and hope for the future.

Educating and Preparing for Future Events:

Lessons Learned: Use the experiences gained during the nuclear event to educate yourself and others about preparedness for future crises. Review what worked well and what could be improved in your response to the situation. Sharing these lessons can help build a more resilient community.

Continuous Preparation: Make mental health a key component of ongoing emergency preparedness. Just as you would stockpile food, water, and medical supplies, consider ways to prepare psychologically for future events. This includes having access to mental health resources, maintaining social connections, and practicing coping strategies regularly.

Dealing with fallout anxiety requires a comprehensive approach that includes immediate coping strategies, long-term mental health care, and a supportive environment. By understanding the sources of anxiety and employing practical techniques to manage stress, you can maintain emotional well-being during a nuclear crisis. Building a supportive community, creating a comfortable living space, and focusing on hope and resilience are key to overcoming the psychological challenges of fallout anxiety. Whether you are caring for yourself or supporting others, these strategies will help you navigate the emotional toll of a nuclear event and emerge stronger on the other side. Remember that mental health is just as important as physical health in a crisis. By prioritizing both, you can ensure that you and your loved ones are better equipped to face the challenges ahead, recover from the trauma, and build a more secure and resilient future.

The Role of Technology in Fallout Shelters

Technology plays a crucial role in enhancing the safety, comfort, and sustainability of fallout shelters. From radiation detection and air filtration to communication systems and renewable energy, modern technology can significantly improve the chances of survival during and after a nuclear event. This chapter explores the various technological tools and systems that can be integrated into fallout shelters to ensure maximum protection, efficiency, and long-term livability.

Radiation Detection and Monitoring

Geiger Counters and Dosimeters:

Geiger Counters: Geiger counters are essential tools for detecting and measuring radiation levels both inside and outside the shelter. These devices provide real-time information about the presence of radioactive particles, allowing you to assess the safety of your environment and make informed decisions about when it's safe to exit the shelter.

Dosimeters: Personal dosimeters track the cumulative radiation dose an individual has received over time. This information is crucial for monitoring long-term exposure and ensuring that individuals do not exceed safe radiation limits. Dosimeters can be worn on the body and are especially useful in scenarios where people need to leave the shelter for short periods.

Radiation Alarms and Alerts:

Radiation Alarms: Some advanced radiation detection systems come equipped with alarms that sound when radiation levels exceed a predetermined threshold. These alarms can provide an immediate warning if radiation levels suddenly rise, allowing you to take protective actions quickly.

Automated Monitoring Systems: Automated systems that continuously monitor radiation levels and send alerts to your phone or other devices can be installed in and around the shelter. These systems provide a constant stream of data, helping you stay informed without the need for manual checks.

Air Filtration and Ventilation Systems

HEPA and Carbon Filters:

HEPA Filters: High-Efficiency Particulate Air (HEPA) filters are designed to remove 99.97% of airborne particles, including radioactive dust, from the air. Integrating HEPA filters into your shelter's ventilation system ensures that the air you breathe is free from harmful contaminants. Regular maintenance and replacement of these filters are necessary to maintain their effectiveness.

Activated Carbon Filters: Activated carbon filters are highly effective at removing chemical contaminants and radioactive iodine from the air. These filters work by adsorbing gases and vapors, making them an essential component of a comprehensive air filtration system.

Ventilation Systems:

Powered Ventilation: A powered ventilation system is crucial for circulating fresh air through the shelter while filtering out radioactive particles. These systems can be powered by electricity, solar energy, or hand-crank generators, ensuring a continuous supply of clean air even during power outages.

Manual Ventilation: In case of power failure, manual ventilation systems such as hand-crank air pumps or gravity-driven air vents can be used to maintain airflow. These systems should be easy to operate and capable of providing adequate ventilation to prevent the buildup of carbon dioxide and other harmful gases.

Communication and Information Systems

Emergency Radios:

Shortwave and HAM Radios: Shortwave and HAM radios are vital for receiving emergency broadcasts, weather updates, and communication from other survivors or authorities. These radios can operate on multiple frequencies, making them versatile tools for staying informed during a crisis.

Crank and Solar-Powered Radios: Crank and solar-powered radios are ideal for fallout shelters, as they do not rely on external power sources. These radios can be powered manually or by sunlight, ensuring that you can receive critical information even if other power sources are unavailable.

Satellite Communication:

Satellite Phones: Satellite phones provide reliable communication in remote or disaster-affected areas where traditional phone networks may be down. These phones can be used to contact emergency services, coordinate with other survivors, or receive updates from government agencies.

Satellite Internet: In some cases, satellite internet systems can be installed in fallout shelters to provide internet access. While this technology may not be available or practical for all shelters, it can be a valuable resource for accessing information, staying in touch with others, and coordinating long-term recovery efforts.

Digital Monitoring and Control Systems:

Smart Shelter Systems: Advanced shelters can be equipped with smart systems that monitor and control various aspects of the shelter environment, such as temperature, humidity, air quality, and security. These systems can be managed via a central control panel or remotely through a smartphone or tablet.

Surveillance Cameras: Surveillance cameras placed around the exterior of the shelter can provide real-time video feeds, allowing you to monitor the area for potential threats or changes in the environment. These systems can be linked to motion detectors and alarms to enhance security.

Power and Energy Solutions

Renewable Energy Sources:

Solar Panels: Solar panels are a sustainable and reliable source of energy for fallout shelters. They can be installed on the shelter's roof or in a nearby area with good sunlight exposure. Solar panels can power essential systems such as lighting, communication devices, and ventilation. Coupled with battery storage, solar energy can provide power even during nighttime or cloudy conditions.

Wind Turbines: Small wind turbines can be used in conjunction with solar panels to generate additional power. These turbines are especially useful in areas with consistent wind patterns and can provide energy when sunlight is limited.

Backup Generators:

Diesel and Gas Generators: Diesel and gas generators provide a reliable backup power source for shelters, especially during extended power outages. Ensure that you have enough fuel stored safely to run the generator for an extended period, and consider using fuel stabilizers to prolong the shelf life of the fuel.

Battery Storage Systems: Battery storage systems are essential for storing energy generated by renewable sources. These systems ensure that you have a continuous power supply even when solar or wind energy is unavailable. Choose high-capacity, deep-cycle batteries designed for long-term use.

Energy Efficiency:

LED Lighting: Use energy-efficient LED lighting throughout your shelter. LEDs consume less power and have a longer lifespan than traditional incandescent bulbs, making them ideal for situations where energy conservation is critical.

Energy Management Systems: An energy management system can help monitor and optimize power usage within the shelter. These systems track energy consumption and adjust power distribution to ensure that critical systems remain operational while minimizing waste.

Water and Food Supply Technologies

Water Filtration and Purification:

UV Water Purifiers: UV water purifiers use ultraviolet light to kill bacteria, viruses, and other pathogens in water. These systems are effective at ensuring that stored or collected water remains safe to drink. UV purifiers are particularly useful in combination with other filtration methods to remove physical contaminants.

Reverse Osmosis Systems: Reverse osmosis systems are highly effective at removing dissolved solids, chemicals, and radioactive particles from water. These systems can be integrated into your shelter's water supply to ensure a continuous supply of clean drinking water.

Food Preservation and Storage:

Dehydrators and Freeze-Dryers: Dehydrators and freeze-dryers are valuable tools for preserving food for long-term storage. These devices remove moisture from food, preventing spoilage and extending shelf life.

Dehydrated and freeze-dried foods are lightweight, require minimal storage space, and retain most of their nutritional value.

Vacuum Sealers: Vacuum sealers are useful for preserving food by removing air from storage bags or containers. This process reduces the risk of spoilage and extends the shelf life of perishable items. Vacuum-sealed foods are protected from moisture, oxygen, and pests, making them ideal for long-term storage.

Automated Hydroponic Systems:

Indoor Gardening: Automated hydroponic systems allow you to grow fresh vegetables and herbs inside your shelter, providing a sustainable source of food. These systems use water-based nutrient solutions instead of soil, making them ideal for enclosed environments. Automated systems can regulate light, water, and nutrients, ensuring optimal growth with minimal manual intervention.

Vertical Farming: If space is limited, vertical farming systems can maximize your food production by growing plants in stacked layers. This technology is particularly effective in shelters with limited floor space, as it allows you to produce a significant amount of food in a small area.

Sanitation and Waste Management

Composting Toilets:

Eco-Friendly Waste Disposal: Composting toilets convert human waste into compost, which can be safely disposed of or used as fertilizer. These toilets do not require water, making them ideal for shelters with limited water supplies. Composting toilets also help reduce odors and minimize the risk of contamination.

Maintenance and Operation: Ensure that composting toilets are properly maintained to prevent issues such as odors or leaks. Regularly empty the compost bin and follow manufacturer guidelines for adding composting materials, such as sawdust or coconut coir, to aid the decomposition process.

Greywater Recycling Systems:

Water Conservation: Greywater recycling systems capture and treat water from sinks, showers, and other non-toilet sources for reuse in tasks such as flushing toilets or watering plants. These systems help conserve water and reduce the demand on your primary water supply.

Filtration and Treatment: Greywater systems typically include filtration and treatment components to remove contaminants before reuse. Ensure that the system is regularly maintained to prevent clogs and ensure the water remains safe for its intended use.

Waste Incinerators:

Waste Disposal: Waste incinerators can be used to safely dispose of non-compostable waste, such as plastic, paper, and hazardous materials. These devices burn waste at high temperatures, reducing it to ash and minimizing the volume of waste that needs to be stored or managed.

Air Filtration: When using an incinerator, ensure that it is equipped with proper air filtration to prevent the release of harmful pollutants. Some incinerators come with built-in filters or scrubbers to clean the exhaust before it is released into the environment. This is particularly important in a confined space like a fallout shelter, where air quality must be carefully managed to ensure the safety and health of its occupants.

Security and Surveillance Systems

Perimeter Security:

Motion Sensors and Alarms: Motion sensors and alarms can be installed around the perimeter of your shelter to detect unauthorized entry or movement. These systems can alert you to potential threats, whether human or animal, and give you time to respond appropriately. Motion-activated lights can also be a deterrent to intruders.

Reinforced Entry Points: Ensure that all entry points to your shelter, including doors and hatches, are reinforced with strong materials like steel and equipped with secure locking mechanisms. Consider installing deadbolts, security bars, and tamper-proof hinges to enhance protection.

Surveillance Cameras:

Exterior Monitoring: Install surveillance cameras outside your shelter to monitor the surrounding area. These cameras can provide real-time video feeds to a central monitoring system, allowing you to keep an eye on your environment without exposing yourself to potential hazards.

Remote Viewing: Some surveillance systems offer remote viewing capabilities, allowing you to access camera feeds from a smartphone or computer. This feature is particularly useful if you need to monitor the shelter's exterior while remaining safely inside.

Internal Security:

Intruder Detection: In addition to exterior security, consider installing internal security measures such as door alarms or surveillance cameras within the shelter itself. These can help you monitor different sections of the shelter, especially if it's large or has multiple rooms.

Safe Rooms: If possible, designate a secure area within the shelter as a "safe room" where you can retreat in case of an internal security breach. This room should have reinforced walls, a secure door, and communication tools to call for help or monitor the situation.

Automation and Smart Technology

Smart Home Integration:

Automated Control Systems: Integrate smart home technology into your fallout shelter to automate various systems, such as lighting, climate control, security, and ventilation. These systems can be controlled via a central hub or remotely through a smartphone, allowing you to manage the shelter's environment efficiently.

Energy Management: Use smart technology to monitor and optimize energy usage within the shelter. Automated systems can prioritize power to essential functions during shortages and reduce energy consumption by adjusting lighting, heating, and other non-essential systems.

Environmental Monitoring:

Temperature and Humidity Control: Automated sensors can continuously monitor temperature and humidity levels within the shelter, making adjustments as needed to maintain a comfortable and safe living environment. These systems can also help prevent mold growth and other issues associated with high humidity.

Air Quality Sensors: Install sensors that detect levels of carbon dioxide, carbon monoxide, and other potentially harmful gases. These sensors can trigger ventilation systems or alarms if air quality falls below safe levels, ensuring that you and your family can take corrective action promptly.

Maintenance Alerts:

Predictive Maintenance: Smart technology can be used to monitor the condition of critical systems within the shelter, such as power generators, air filtration units, and water purification systems. These systems can provide alerts when maintenance is required or when a component is nearing the end of its lifespan, helping to prevent breakdowns.

Automated Reporting: Automated systems can generate reports on the status of various shelter components, such as energy usage, water supply levels, and filter efficiency. These reports can be invaluable for long-term planning and ensuring that all systems are functioning optimally.

Preparing for Long-Term Habitation

Sustainable Living Solutions:

Aquaponics and Hydroponics: Consider incorporating aquaponics (a system that combines fish farming with plant cultivation) into your shelter's design. This system can provide a renewable source of both protein and fresh vegetables, creating a self-sustaining food supply. Hydroponics, mentioned earlier, can be supplemented with fish in an aquaponics system to create a more diverse and efficient food production setup.

Renewable Energy Storage: Invest in advanced energy storage solutions, such as lithium-ion or flow batteries, to store energy generated by solar panels or wind turbines. These systems can provide a reliable power source for extended periods, even when renewable energy input is low.

Psychological Comfort:

Entertainment Systems: Include entertainment options such as a digital library, games, music, and movies to keep morale high during long periods in the shelter. Consider offline media options, such as DVDs or books, in case internet access is unavailable.

Communication Tools: Maintain communication with the outside world through secure channels, even if only sporadically. The ability to connect with others, receive news, and share experiences can significantly alleviate the psychological strain of isolation.

Health and Wellness:

Exercise Equipment: Space-permitting, include exercise equipment such as resistance bands, hand weights, or a stationary bike to help maintain physical health. Regular exercise is also crucial for mental well-being, reducing stress, and maintaining a sense of normalcy.

Medical Technology: Consider equipping your shelter with basic medical devices, such as a defibrillator, blood pressure monitor, and first aid supplies, to manage health emergencies. Telemedicine options, if available, can provide remote medical consultations.

Technology plays a pivotal role in transforming a fallout shelter from a basic refuge into a fully equipped, sustainable living space capable of supporting long-term habitation. By integrating advanced systems for radiation detection, air filtration, communication, energy management, and security, you can significantly enhance the safety, comfort, and efficiency of your shelter.

As you prepare your fallout shelter, consider the specific challenges you may face and how technology can address these challenges. From maintaining air quality and power to ensuring a reliable food and water supply, the right technological solutions can make all the difference in surviving and thriving during a nuclear event. Remember that technology also includes the ability to maintain your mental health and well-being, which is just as important as physical survival.

DIY Fallout Shelter: Building on a Budget

Building a fallout shelter on a budget is a practical and achievable goal for many people, especially when considering the potential need for protection against nuclear fallout or other disasters. While constructing a high-tech, state-of-the-art shelter may be out of reach for some, there are numerous ways to create an effective and functional fallout shelter without breaking the bank. This chapter will guide you through the steps of building a DIY fallout shelter on a budget, focusing on affordable materials, cost-effective construction methods, and essential features to ensure your safety and comfort.

Planning Your DIY Fallout Shelter

Assessing Your Needs and Budget:

Determine the Shelter's Purpose: Before starting construction, define the primary purpose of your shelter. Is it meant to provide short-term protection during a nuclear event, or is it intended for long-term habitation? Your answer will influence the size, location, and features of the shelter.

Set a Budget: Determine how much you can afford to spend on the shelter. Consider all potential costs, including materials, tools, labor (if not doing it all yourself), and any permits or fees required by local authorities. Prioritize essential features like structural integrity, radiation shielding, and ventilation, and allocate funds accordingly.

Location and Design Considerations:

Choosing the Right Location: The location of your shelter is crucial for its effectiveness. Ideally, the shelter should be underground or partially buried to take advantage of natural earth shielding. If digging is not possible, consider building the shelter in a basement or converting an existing room into a shelter.

Simple and Functional Design: Opt for a simple and functional design that prioritizes safety and usability. A rectangular or square floor plan is often the most cost-effective and easiest to construct. Focus on essential features such as thick walls, a secure door, and adequate ventilation.

Obtaining Permits and Approvals:

Check Local Regulations: Before beginning construction, check local building codes and regulations to ensure that your shelter complies with legal requirements. Some areas may require permits for underground construction, and it's important to address these issues upfront to avoid fines or complications later.

Consulting Experts: If you're unsure about any aspect of the construction process, consider consulting with a structural engineer or other experts who can provide guidance on safety and best practices. While this may involve some additional cost, it can save money and prevent costly mistakes in the long run.

Affordable Materials for Building Your Shelter

Using Readily Available Materials:

Concrete Blocks: Concrete blocks are an affordable and effective material for building the walls of your shelter. They provide good radiation shielding and are easy to stack and mortar together. If possible, reinforce the walls with rebar and fill the blocks with concrete to enhance their strength.

Earth and Sand: If you're building an underground or partially buried shelter, use the excavated earth or sand as part of your construction. Earth provides excellent radiation shielding, and sand can be used to fill bags or create barriers around the shelter.

Timber and Plywood: Timber and plywood are cost-effective materials that can be used for framing, flooring, and interior walls. While they do not provide much radiation shielding on their own, they are useful for creating a structure that can then be reinforced with other materials.

Recycled and Salvaged Materials:

Reclaimed Wood and Metal: Look for reclaimed wood, metal beams, and other materials from demolition sites, salvage yards, or classified ads. These materials can be significantly cheaper than new ones and can still provide excellent structural support.

Shipping Containers: Shipping containers are a popular and affordable option for DIY shelters. They are durable, secure, and relatively easy to bury or reinforce. If using a shipping container, add interior insulation and extra radiation shielding to make it more suitable for long-term habitation.

Cost-Effective Radiation Shielding:

Sandbags: Sandbags are a low-cost and highly effective way to add radiation shielding to your shelter. They can be stacked around the exterior walls or even used to create a protective barrier on the roof. Sand is dense enough to reduce radiation exposure significantly, and sandbags are easy to source and use.

Water Barrels: Water is an excellent radiation shield, and storing water in barrels inside your shelter can serve a dual purpose. The barrels provide radiation protection while also ensuring a supply of drinking water. Place them strategically around the shelter to maximize their shielding effect.

Construction Techniques on a Budget

Building an Underground Shelter:

Excavation: If you're building an underground shelter, excavation is one of the most labor-intensive parts of the process. To save money, consider doing the digging yourself or hiring local labor at a lower cost. You can also rent digging equipment, but ensure that you're familiar with its operation to avoid costly mistakes.

Simple Roof Design: For the roof of an underground shelter, use a simple design with reinforced concrete or thick wooden beams covered with earth. Adding a layer of plastic sheeting or a waterproof membrane between the roof and the earth can help prevent moisture infiltration.

Above-Ground Shelters:

Retrofitting Existing Structures: One of the most cost-effective ways to build a fallout shelter is to retrofit an existing structure, such as a basement, garage, or outbuilding. Strengthen the walls with concrete blocks or sandbags and seal any windows or doors to prevent radiation from entering.

Earthen Berms: For above-ground shelters, consider building earthen berms around the structure. These are mounds of earth piled against the walls, providing additional radiation shielding and insulation. This method is particularly useful if you cannot bury the shelter entirely.

Door and Ventilation Solutions:

Reinforced Doors: The door is one of the most critical components of your shelter. Use a solid wood or metal door reinforced with steel plates or sandbags. Ensure that it has a secure locking mechanism and a tight seal to prevent radiation from entering.

DIY Ventilation: For ventilation, you can create a simple air exchange system using PVC pipes, hand-cranked fans, and HEPA filters. Ensure that the ventilation system includes a way to filter out radioactive particles, and consider adding a manual backup in case of power failure.

Essential Features and Add-Ons

Basic Amenities:

Water Storage and Purification: Include a reliable system for storing and purifying water. Large, food-grade barrels or containers are affordable options for water storage. Add a gravity-fed water filter or portable purification tablets to ensure that your water supply remains safe to drink.

Food Storage: Stock your shelter with non-perishable food items, such as canned goods, rice, beans, and freeze-dried meals. Store food in airtight containers to protect it from moisture and pests. A small propane or alcohol stove can be used for cooking, provided you have adequate ventilation.

Sanitation Solutions:

Portable Toilet: A portable camping toilet or bucket system lined with heavy-duty plastic bags can serve as a simple and affordable sanitation solution. Ensure that you have a supply of waste bags, disinfectants, and absorbent materials like sawdust or cat litter to manage waste.

Waste Disposal: Plan for waste disposal by setting up a designated area within the shelter for storing sealed waste bags. If possible, create a small, vented compartment for waste storage to keep it separate from living areas and reduce odors.

Power and Lighting:

Battery-Powered Lights: Battery-powered LED lights are an affordable and energy-efficient option for shelter lighting. Stock up on extra batteries or rechargeable batteries with a solar charger to ensure a consistent power source.

Backup Power: If your budget allows, consider investing in a small, portable generator or a solar power system to run essential devices like radios, ventilation fans, and lights. Solar chargers can also be used to recharge batteries and small electronics.

Testing and Maintenance on a Budget

Regular Inspections:

Check Structural Integrity: Even if you're building on a budget, regular inspections are crucial to ensure that your shelter remains safe and functional. Look for signs of wear, such as cracks in walls, moisture buildup, or rust on metal components, and make repairs as needed.

Testing Systems: Periodically test all systems in your shelter, including ventilation, water storage, and lighting, to ensure they are working correctly. Replace any failing components promptly to avoid issues during an emergency.

DIY Maintenance:

Simple Repairs: Learn how to perform basic maintenance tasks yourself, such as sealing cracks, replacing filters, or repairing leaks. Having the skills to do your own maintenance can save money and ensure that your shelter is always ready for use.

Emergency Supplies: Keep a supply of basic tools, spare parts, and repair materials in your shelter so that you can quickly address any issues that arise. Items like duct tape, caulking, plastic sheeting, and a multi-tool can be invaluable during an emergency.

Final Tips for Building on a Budget

Prioritize Safety:

Don't Skimp on Essentials: While it's important to save money, never compromise on the essential safety features of your shelter, such as structural integrity, radiation shielding, and ventilation. These elements are critical for your survival and should be the focus of your budget.

Scavenge and Reuse:

Use What You Have: Before purchasing new materials, look around your home, garage, or property for items that can be repurposed for the shelter. Old doors, metal sheets, lumber, and even furniture can be modified to serve as part of your shelter.

Community Resources: Check local community groups, online marketplaces, or salvage yards for free or low-cost building materials. Sometimes, you can find quality materials that others are giving away, significantly reducing your costs

Build in Stages:

Start Small: If your budget is tight, consider building your shelter in stages. Start with the most critical components, such as the basic structure and radiation shielding, and add features like ventilation systems, enhanced security, or comfort amenities as funds become available. This phased approach allows you to prioritize safety while spreading out the costs over time.

Upgrade Over Time: As your financial situation improves or as you find additional materials, you can continue to enhance your shelter. Upgrades could include better insulation, more advanced filtration systems, or the addition of renewable energy sources.

DIY and Barter:

Do-It-Yourself: Labor costs can quickly add up, so doing as much of the work yourself as possible is a great way to save money. There are plenty of online resources, including videos, forums, and guides, that can teach you the necessary skills, from basic construction techniques to installing ventilation systems.

Barter Skills: If you lack certain skills, consider bartering with others in your community. Perhaps you're good at electrical work, while a neighbor excels at carpentry. Swapping services can help you both save money while completing your shelter.

Stay Informed:

Research and Learning: Stay informed about the latest DIY techniques, affordable building materials, and emergency preparedness strategies. The more you know, the better equipped you'll be to make smart decisions and find cost-effective solutions.

Practical Examples of DIY Fallout Shelters

Example 1: Basement Conversion:

Overview: Converting a basement into a fallout shelter is one of the most practical and budget-friendly options for those who already have a suitable space. Reinforce the walls with concrete blocks or sandbags, seal any windows or doors, and add a simple ventilation system.

Cost Estimate: Depending on the size of the basement and the materials used, this project could cost anywhere from a few hundred to a couple of thousand dollars. Using salvaged or recycled materials can significantly reduce costs.

Example 2: Shipping Container Shelter:

Overview: Shipping containers are robust, readily available, and relatively inexpensive. They can be buried or surrounded by earthen berms for additional radiation protection. Inside, add insulation, a ventilation system, and essential supplies.

Cost Estimate: A used shipping container can cost between $1,500 and $4,500, depending on its condition and size. Additional costs include insulation, reinforcing, and any customizations needed to convert the container into a liveable space.

Example 3: Backyard Underground Shelter:

Overview: For those with the space and ability to dig, a simple underground shelter can be constructed using concrete blocks or timber for walls and a concrete slab or earth-covered roof. This type of shelter provides excellent protection at a relatively low cost.

Cost Estimate: Excavation costs can vary widely depending on whether you do the work yourself or hire someone. Overall, this project could range from $2,000 to $10,000, depending on the depth, size, and materials used.

Preparing for Emergency Use

Stocking the Shelter:

Essential Supplies: Ensure that your shelter is stocked with essential supplies, including food, water, medical supplies, and tools. Focus on non-perishable items and consider long-term storage solutions to keep these supplies safe and ready for use.

Personal Items: Add personal items that can help maintain morale during a crisis, such as books, games, or comfort items like blankets and pillows. These small touches can make a significant difference during long periods in the shelter.

Drills and Practice:

Regular Drills: Practice entering the shelter quickly and efficiently, especially with family members. Familiarize everyone with the location of emergency supplies, the operation of ventilation systems, and the use of any communication devices.

Reviewing Procedures: Periodically review and update your emergency procedures. Make sure everyone knows what to do in various scenarios, from a sudden need to shelter in place to a long-term stay.

Emergency Repairs:

Quick Fixes: Keep materials on hand for emergency repairs, such as duct tape, patching materials, and basic tools. Being able to quickly address issues like leaks or damaged walls can be critical during an extended stay in the shelter.

Maintenance Kit: Prepare a maintenance kit that includes spare parts for essential systems like ventilation and lighting. Regularly check that everything in the kit is in good condition and easily accessible.

Adapting and Expanding Your Shelter

Adapting to Changing Needs:

Flexible Design: Design your shelter with flexibility in mind. For example, include space that can be used for multiple purposes, such as storage that can also serve as sleeping areas. As your needs evolve, your shelter can adapt accordingly.

Room for Growth: If possible, leave room for future expansion. You may start with a small, basic shelter and gradually expand it as resources allow. This could include adding additional storage, sleeping areas, or improved systems.

Long-Term Planning:

Sustainability: Plan for long-term sustainability by considering renewable energy options, such as adding solar panels or wind turbines as funds become available. Also, think about ways to grow food or collect water within or near the shelter.

Community Support: As you develop your shelter, consider how it fits into a broader community preparedness plan. Working with neighbors or a local preparedness group can provide mutual support and resources, making your efforts more effective and sustainable.

Building a DIY fallout shelter on a budget is a feasible and rewarding project that can provide critical protection in the event of a nuclear disaster or other emergencies. By carefully planning, using affordable materials, and employing cost-effective construction techniques, you can create a shelter that meets your safety needs without requiring a significant financial investment. While it's essential to prioritize safety and functionality, remember that even on a limited budget, there are creative ways to enhance your shelter's comfort and sustainability. With patience, resourcefulness, and a focus on the most critical features, you can build a reliable fallout shelter that offers peace of mind and security for you and your loved ones.

Advanced Engineering for Fallout Shelters

Building a fallout shelter that offers maximum protection, sustainability, and comfort requires the application of advanced engineering principles. While basic shelters can be effective for short-term protection, advanced engineering techniques can elevate a shelter's capabilities, making it suitable for long-term habitation in the most challenging conditions. This chapter will explore the various advanced engineering concepts and technologies that can be integrated into fallout shelters to enhance their durability, safety, and liveability.

Structural Engineering for Enhanced Durability

Seismic-Resistant Design:

Reinforced Concrete: In areas prone to earthquakes, it's essential to design a shelter that can withstand seismic activity. Reinforced concrete is a robust material that can be engineered to resist both lateral and vertical forces caused by seismic events. Reinforcing the concrete with steel rebar, designed in a grid pattern, increases the structure's ability to absorb and dissipate energy without collapsing.

Base Isolation Systems: Base isolation involves placing flexible bearings or isolators between the shelter's foundation and the structure itself. This technology allows the shelter to move independently of the ground during an earthquake, significantly reducing the impact of seismic forces on the structure.

Blast-Resistant Construction:

Shock-Absorbing Materials: Incorporate shock-absorbing materials, such as laminated rubber or elastomeric bearings, into the walls and foundation to mitigate the effects of a nearby blast. These materials help absorb and dissipate the energy from shockwaves, reducing the likelihood of structural damage.

Honeycomb or Composite Structures: Honeycomb structures, made from materials like aluminum or steel, provide excellent blast resistance due to their ability to deform under pressure and absorb energy. Composite materials, such as carbon fiber reinforced polymers, offer high strength-to-weight ratios and can be used to reinforce critical areas of the shelter.

Underground Shelter Stability:

Geotechnical Engineering: Understanding the soil composition and geological conditions at the shelter site is crucial for designing a stable underground structure. Conduct soil testing to determine the load-bearing capacity and potential for soil liquefaction. Based on these results, engineers can design appropriate foundation systems, such as deep foundations, piles, or retaining walls, to ensure the shelter remains stable over time.

Slope Stabilization: If the shelter is built into a hillside or slope, it's essential to stabilize the surrounding earth to prevent landslides or erosion. Techniques such as retaining walls, geotextiles, and soil nailing can be employed to reinforce the slope and protect the shelter.

Advanced Radiation Shielding Techniques

Multi-Layered Shielding Systems:

Layered Shielding: Use a multi-layered approach to radiation shielding by combining materials with different properties. For example, a layer of lead (which blocks gamma rays) can be followed by a layer of concrete (which provides structural support and absorbs neutron radiation), and finally, a layer of water or polyethylene (which further absorbs neutrons and beta particles).

Active Radiation Shielding: While still largely experimental, active radiation shielding involves creating electromagnetic fields to deflect or reduce radiation. This technology is inspired by research in space travel and could potentially be adapted for terrestrial shelters in the future.

Shielded Ventilation Systems:

HEPA and Carbon Filters: Incorporate high-efficiency particulate air (HEPA) filters in the ventilation system to remove radioactive particles from the air. These filters can be paired with activated carbon filters to absorb radioactive gases like iodine-131.

Baffle Systems: A baffle system in the air intake and exhaust vents can trap radioactive particles and prevent them from entering the shelter. This system involves a series of angled plates or walls that force the air to change direction multiple times, causing heavier particles to fall out of the airflow.

Radiation-Resistant Materials:

Radiation-Absorbing Concrete: Engineers can design concrete with additives like boron or barite to enhance its ability to absorb neutron radiation. This type of concrete is often used in nuclear power plants and can be adapted for fallout shelters to provide superior radiation protection.

Lead-Glass Windows: If your shelter includes windows or observation ports, consider using lead-glass, which contains a high concentration of lead oxide. This material provides excellent visibility while offering substantial protection against gamma radiation.

Advanced Air Filtration and Ventilation

Positive Pressure Ventilation:

Maintaining Positive Pressure: Positive pressure ventilation systems are designed to keep the air pressure inside the shelter slightly higher than the outside. This prevents contaminated air from seeping in through small cracks or gaps, as air flows out rather than in. A combination of air pumps and HEPA filters ensures that the air entering the shelter is clean and safe.

Backup Power Systems: To maintain positive pressure consistently, it's vital to have reliable backup power systems, such as generators or battery storage, to keep the ventilation system running even during power outages.

Automated Environmental Controls:

Climate Control Systems: Advanced climate control systems can regulate temperature and humidity within the shelter automatically. These systems use sensors to monitor the environment and adjust heating, cooling, and dehumidification systems as needed, ensuring a comfortable living space regardless of external conditions.

Air Quality Sensors: Install sensors that continuously monitor air quality within the shelter, detecting levels of carbon dioxide, carbon monoxide, and other potentially harmful gases. These sensors can trigger ventilation adjustments or alarms if air quality falls below safe levels.

Oxygen Generation and Recycling:

Oxygen Concentrators: In long-term scenarios where fresh air is limited, oxygen concentrators can be used to extract oxygen from the surrounding air, concentrating it for use in the shelter. These devices are especially useful in underground shelters where air circulation may be restricted.

Electrolysis Systems: Electrolysis systems can generate oxygen by splitting water molecules into hydrogen and oxygen. This method provides a renewable source of oxygen, particularly useful in sealed environments, although it requires a consistent power supply.

Water and Waste Management Innovations

Water Recycling and Filtration:

Greywater Recycling Systems: Advanced greywater recycling systems treat and reuse water from sinks, showers, and other non-toilet sources for flushing toilets or watering plants. These systems use multi-stage filtration processes, including biological, mechanical, and chemical treatments, to ensure the recycled water is safe for its intended use.

Desalination Units: If your shelter is located near a seawater source, consider installing a desalination unit to convert seawater into potable water. Reverse osmosis is the most common desalination method, but it requires significant energy, so pairing it with renewable energy sources is advisable.

Composting and Incineration Toilets:

Advanced Composting Toilets: Modern composting toilets use aerobic processes to break down waste into compost efficiently. These systems are designed to minimize odor, require minimal maintenance, and can operate off-grid, making them ideal for long-term shelter use.

Waste Incineration Systems: Incineration toilets use electricity or gas to burn waste at high temperatures, reducing it to sterile ash. This method is effective in eliminating pathogens and reducing waste volume, but it requires a consistent energy source and proper ventilation.

Automated Waste Management:

Smart Waste Disposal: Automated waste disposal systems can sort and process different types of waste, directing compostable material to composting units and non-organic waste to incineration or storage. These systems reduce the need for manual handling of waste, improving hygiene and efficiency.

Bio-solid Processing: For larger shelters, consider bio-solid processing systems that convert human waste into safe, reusable by-products such as fertilizer. These systems are complex and require careful management but offer a sustainable solution for long-term waste management.

Advanced Power Generation and Storage

Hybrid Renewable Energy Systems:

Solar-Wind Hybrid Systems: Combine solar panels with wind turbines to create a hybrid renewable energy system that can generate power continuously, even when one source is unavailable. This approach ensures a more reliable power supply and can be tailored to the specific environmental conditions of your shelter's location.

Geothermal Energy: If your shelter is located in an area with geothermal activity, consider harnessing this energy source. Geothermal systems use the heat from the Earth to generate electricity or provide heating, offering a stable and consistent energy supply.

Energy Storage Solutions:

Advanced Battery Storage: Lithium-ion batteries are commonly used for energy storage, but newer technologies such as solid-state batteries or flow batteries offer higher efficiency, longer life, and greater safety. These batteries store excess energy generated by renewable sources for use during periods of low generation.

Compressed Air Energy Storage (CAES): CAES systems store energy by compressing air in underground caverns or tanks, which can then be released to drive turbines and generate electricity when needed. This technology is scalable and can provide long-term energy storage for large shelters.

Smart Grid Integration:

Micro grids: A micro grid is a localized energy grid that can operate independently from the main power grid. By integrating renewable energy sources, energy storage, and smart distribution systems, a micro grid can provide reliable power to your shelter even during widespread grid failures.

Energy Management Systems: Smart energy management systems monitor and control the use of electricity within the shelter, optimizing power distribution, and reducing waste. These systems can prioritize critical loads and automatically switch to backup power sources when necessary.

Smart Technology and Automation

Integrated Control Systems:

Shelter Automation: Advanced shelters can be equipped with integrated control systems that allow you to monitor and manage all aspects of the shelter's environment, from temperature and lighting to security and power usage, from a single interface. These systems can be controlled via touchscreens or remotely using a smartphone or tablet.

AI and Machine Learning: Incorporating AI and machine learning into your shelter's automation systems can enhance their efficiency and adaptability. AI can analyze data from various sensors to optimize energy usage, predict equipment maintenance needs, and adjust environmental controls based on changing conditions. Machine learning algorithms can learn your preferences and habits over time, further personalizing the shelter's environment to improve comfort and efficiency.

Robotics and Automation:

Automated Maintenance Robots: Robots equipped with AI can perform routine maintenance tasks within the shelter, such as cleaning, inspecting for structural integrity, and monitoring equipment. These robots reduce the need for manual labor, especially in hazardous conditions, and can alert you to potential issues before they become critical.

Robotic Security Systems: Advanced shelters can incorporate robotic security systems that patrol the perimeter or monitor specific areas. These robots can be equipped with cameras, motion sensors, and other detection equipment to provide real-time surveillance and respond to potential threats automatically.

3D Printing for On-Demand Manufacturing:

3D Printing of Replacement Parts: 3D printers can be used within the shelter to manufacture replacement parts for equipment or tools as needed. This capability is especially valuable in long-term scenarios where external supplies may be limited. With the right materials, 3D printers can produce durable, high-quality components.

Construction and Repairs: Large-scale 3D printing technology can even be used to construct additional shelter spaces or reinforce existing structures. 3D-printed concrete, for example, allows for rapid construction of walls or barriers, providing an additional layer of protection.

Virtual Reality (VR) and Augmented Reality (AR) for Training and Maintenance:

Virtual Training Environments: VR can be used to simulate various emergency scenarios, allowing you and your family to practice responses in a safe, controlled environment. This type of training is invaluable for preparing for real-world crises without the associated risks.

Augmented Reality Maintenance: AR devices can assist in maintenance tasks by overlaying instructions or diagnostic information onto the physical environment. For example, wearing AR glasses, you could see the exact location of wiring or pipes within a wall, making repairs more accurate and efficient.

Advanced Security and Surveillance Systems

Integrated Security Networks:

Layered Security: Advanced shelters can employ a layered security approach, integrating physical barriers, electronic surveillance, and cybersecurity measures. Physical barriers include reinforced doors, bulletproof glass, and secure entry points, while electronic systems might involve biometric access controls and motion-detecting cameras.

Cybersecurity for Smart Systems: As shelters become more automated and connected, they also become more vulnerable to cyberattacks. Implement robust cybersecurity measures to protect your shelter's systems, including firewalls, encryption, and intrusion detection systems.

Biometric Access Control:

Fingerprint Scanners: Fingerprint scanners can be installed at entry points to ensure that only authorized individuals can access the shelter. This technology provides a high level of security while allowing for quick and convenient access.

Retina and Facial Recognition: For even greater security, consider integrating retina scanners or facial recognition systems. These technologies can be combined with other biometric systems for multi-factor authentication, reducing the risk of unauthorized entry.

Advanced Surveillance Technology:

Drones for Perimeter Surveillance: Drones equipped with cameras and sensors can patrol the area around your shelter, providing aerial surveillance and real-time video feeds. These drones can be programmed to follow specific routes or respond to detected movement, enhancing your ability to monitor the environment without exposing yourself to potential threats.

Thermal and Night Vision Cameras: Install thermal imaging and night vision cameras to monitor your shelter's surroundings in low-light conditions or through obstructions like fog or smoke. These cameras can detect heat signatures from humans or animals, providing an additional layer of security.

Psychological and Environmental Design

Biophilic Design Principles:

Incorporating Nature: Biophilic design focuses on incorporating elements of nature into built environments to improve mental well-being. In a fallout shelter, this could mean adding plants that can thrive with artificial light, using natural materials like wood and stone in the interior design, or even integrating water features that create a soothing atmosphere.

Simulated Natural Light: Use advanced lighting systems that mimic natural sunlight to regulate circadian rhythms and improve mood. LED lights that change color and intensity throughout the day can help create a more natural living environment in a shelter that lacks access to natural light.

Stress-Reduction Techniques:

Noise Control: Design the shelter to minimize noise from mechanical systems, ventilation, or other occupants. Soundproofing materials and white noise machines can be used to create a quieter, more peaceful environment, reducing stress and enhancing sleep quality.

Private Spaces: Even in a compact shelter, it's important to design areas where individuals can have privacy. These spaces can be used for meditation, reading, or simply taking a break from the group, which is crucial for mental health during extended stays.

Environmental Enrichment:

Customizable Living Spaces: Allow for customizable spaces within the shelter, where occupants can personalize their surroundings with art, photos, or other personal items. This personalization can help make the environment feel more like home, which is important for maintaining morale.

Entertainment Systems: Include advanced entertainment systems, such as VR headsets or projection systems, to provide immersive experiences that can serve as a mental escape. Access to movies, games, and other media can significantly improve quality of life during long-term confinement.

Sustainability and Self-Sufficiency

Closed-Loop Systems:

Water Recycling: Design a closed-loop water system that recycles greywater for use in toilets, irrigation, or other non-potable applications. Advanced filtration systems can purify this water to a level safe for human consumption if necessary.

Waste-to-Energy Conversion: Integrate waste-to-energy systems that convert organic waste into biogas, which can be used for cooking or heating. This reduces waste volume while providing a renewable energy source, contributing to the shelter's self-sufficiency.

Advanced Agricultural Systems:

Aquaponics: Aquaponics combines fish farming with hydroponics, creating a symbiotic environment where fish waste provides nutrients for plants, and the plants help filter the water for the fish. This system can produce both vegetables and protein, making it ideal for long-term food production in a shelter.

Vertical Farming: Use vertical farming techniques to maximize food production in a limited space. Automated lighting, watering, and nutrient delivery systems ensure optimal growth conditions, allowing for the continuous production of fresh vegetables and herbs.

Renewable Energy Integration:

Advanced Solar Solutions: Incorporate high-efficiency solar panels that can generate power even in low-light conditions. Paired with energy storage systems, these panels can provide a consistent energy supply for essential systems.

Microbial Fuel Cells: An emerging technology, microbial fuel cells generate electricity from organic matter, such as waste or soil, through the action of microorganisms. This technology could be integrated into a shelter to provide a small but continuous power source from organic waste or other materials.

Advanced engineering for fallout shelters goes beyond basic survival, aiming to create an environment that is safe, sustainable, and comfortable for long-term habitation. By integrating cutting-edge technologies and innovative design principles, it's possible to build a shelter that not only protects against the immediate dangers of a nuclear event but also supports a high quality of life during extended stays.

From structural enhancements that improve durability and blast resistance to advanced systems for radiation shielding, air filtration, and energy management, these engineering solutions can transform a fallout shelter into a self-sufficient, resilient refuge. Additionally, incorporating smart technology, automation, and psychological design considerations ensures that the shelter remains liveable and conducive to mental and physical well-being.

Ultimately, while building an advanced fallout shelter requires significant investment and expertise, the benefits in terms of safety, security, and comfort are unparalleled. For those committed to preparing for worst-case scenarios, advanced engineering offers the tools and techniques needed to create a sanctuary capable of withstanding the challenges of a post-nuclear world.

Understanding International Fallout Safety Protocols

In a world where the threat of nuclear conflict remains a concern, understanding international fallout safety protocols is crucial for ensuring the safety and preparedness of populations worldwide. These protocols are developed by governments, international organizations, and safety agencies to mitigate the effects of nuclear fallout and protect civilians during and after a nuclear event. This chapter will explore the key international fallout safety protocols, the organizations responsible for their development and implementation, and how these guidelines can be applied at both national and local levels.

The Role of International Organizations

United Nations and the International Atomic Energy Agency (IAEA):

IAEA's Role: The International Atomic Energy Agency (IAEA) is a key organization within the United Nations system responsible for promoting the safe and peaceful use of nuclear energy. It plays a critical role in establishing global safety standards for radiation protection, nuclear security, and emergency preparedness.

Conventions and Guidelines: The IAEA develops and promotes various conventions, guidelines, and safety standards that member states are encouraged to adopt. These include the Convention on Early Notification of a Nuclear Accident and the Convention on Assistance in the Case of a Nuclear Accident or Radiological Emergency. These documents outline procedures for international cooperation, information sharing, and emergency response during a nuclear event.

World Health Organization (WHO):

Public Health Preparedness: The World Health Organization (WHO) is involved in the global health response to nuclear emergencies. It provides guidance on protecting public health, including protocols for radiation exposure, medical treatment of radiation sickness, and managing contaminated food and water supplies.

International Health Regulations (IHR): The WHO's International Health Regulations provide a framework for coordinating international public health responses to emergencies, including nuclear events. The IHR ensures that countries have the capacity to detect, assess, report, and respond to public health risks, including those posed by radioactive contamination.

International Civil Defense Organization (ICDO):

Civil Defense Measures: The ICDO assists member states in developing and implementing civil defense measures, including those related to nuclear fallout. It provides training, resources, and technical assistance to enhance national and regional preparedness for nuclear emergencies.

Global Safety Frameworks: The ICDO collaborates with other international organizations to create global frameworks for civil protection, focusing on harmonizing safety protocols, standardizing emergency responses, and ensuring that countries are prepared to protect their populations in the event of a nuclear disaster.

Key International Safety Protocols

The Basic Safety Standards (BSS):

Radiation Protection Guidelines: The IAEA's Basic Safety Standards (BSS) provide comprehensive guidelines for protecting people and the environment from harmful effects of ionizing radiation. These standards are widely adopted by countries to regulate the use of nuclear energy, medical radiation, and radioactive materials.

Principles of Radiation Protection: The BSS is based on three fundamental principles: justification (ensuring that any exposure to radiation has more benefits than risks), optimization (keeping radiation doses as low as reasonably achievable), and dose limitation (ensuring that no individual is exposed to radiation beyond prescribed safety limits).

The Joint Radiation Emergency Management Plan of the International Organizations (JPLAN):

Coordinated Emergency Response: The JPLAN, developed by the IAEA and other international organizations, provides a coordinated approach to managing radiation emergencies. It outlines roles and responsibilities for international organizations in responding to nuclear accidents and incidents, ensuring a unified and effective response.

Information Exchange: The JPLAN emphasizes the importance of timely and accurate information exchange between countries and international organizations during a nuclear emergency. This helps to minimize the impact of radiation exposure on populations and facilitates a coordinated global response.

The International Nuclear and Radiological Event Scale (INES):

Classifying Nuclear Incidents: The INES is a tool developed by the IAEA for communicating the safety significance of nuclear and radiological events to the public. It categorizes incidents on a scale from 1 (anomaly) to 7 (major accident) based on their impact on people, the environment, and the facility involved.

Global Communication: The INES scale helps ensure consistent communication about the severity of nuclear incidents, enabling governments and organizations to implement appropriate safety measures based on the level of risk.

National Implementation of International Protocols

Adopting International Standards:

National Regulatory Bodies: Countries typically have national regulatory bodies responsible for adopting and enforcing international safety protocols. These agencies, such as the Nuclear Regulatory Commission (NRC) in the United States or the Office for Nuclear Regulation (ONR) in the United Kingdom, play a crucial role in integrating global standards into national regulations.

Customizing Protocols: While international protocols provide a framework, national authorities often customize them to address specific local conditions, such as geographic vulnerabilities, population density, and available resources. This ensures that safety measures are tailored to the unique risks faced by each country.

Public Awareness and Education:

Government Initiatives: Governments are responsible for raising public awareness about nuclear safety protocols and ensuring that citizens know how to respond in the event of a nuclear emergency. This may involve public information campaigns, educational programs, and regular emergency drills.

Community Engagement: Engaging communities in preparedness activities is vital for the effective implementation of safety protocols. Governments may work with local authorities, non-governmental organizations (NGOs), and community groups to provide training, resources, and support for individuals and families.

Emergency Preparedness Plans:

National Emergency Plans: Countries develop national emergency preparedness plans based on international safety protocols. These plans outline the procedures for evacuations, sheltering, medical response, and public communication during a nuclear event. They also include provisions for coordinating with international organizations and neighboring countries.

Local and Regional Plans: Local and regional governments are responsible for implementing the national emergency plans at the community level. This includes identifying safe zones, establishing communication networks, and ensuring that local emergency services are equipped to handle a nuclear incident.

Challenges in International Fallout Safety

Coordination between Countries:

Cross-Border Challenges: Nuclear fallout does not respect national borders, making international coordination essential. However, differences in regulatory frameworks, communication systems, and resources can complicate cross-border responses. Developing standardized protocols and fostering cooperation between neighboring countries is critical to addressing these challenges.

Language and Cultural Barriers: Effective communication during an international nuclear emergency can be hindered by language and cultural differences. Ensuring that safety information is available in multiple languages and is culturally appropriate is important for protecting diverse populations.

Resource Disparities:

Developed vs. Developing Countries: Resource disparities between developed and developing countries can impact the implementation of international safety protocols. While wealthier nations may have the infrastructure and technology to meet global standards, poorer countries may struggle to achieve the same level of preparedness. International assistance and capacity-building programs are necessary to bridge these gaps.

Access to Technology: Advanced technologies for radiation detection, medical treatment, and emergency communication are not equally available worldwide. Ensuring equitable access to these technologies is a key challenge in global fallout safety.

Compliance and Enforcement:

Voluntary Nature of Protocols: Many international safety protocols are voluntary, relying on countries to adopt and enforce them within their borders. This can lead to inconsistencies in compliance, with some nations fully embracing the guidelines while others may not prioritize them.

Monitoring and Accountability: Ensuring that countries adhere to international safety standards requires effective monitoring and accountability mechanisms. The IAEA and other organizations play a role in assessing compliance, but political and diplomatic factors can influence the enforcement of these protocols.

Applying International Protocols Locally

Community Preparedness Initiatives:

Local Drills and Exercises: Conducting regular drills and exercises at the community level helps to ensure that international safety protocols are effectively implemented. These activities should simulate various nuclear scenarios, allowing local authorities and residents to practice their response and identify areas for improvement.

Public Education Programs: Local governments can adapt international safety guidelines into public education programs that teach citizens about the basics of radiation, how to use protective measures like iodine tablets, and the importance of sheltering in place during a nuclear event.

Integrating Technology in Local Responses:

Radiation Detection Networks: Establishing local networks of radiation detection equipment, connected to national and international monitoring systems, can provide real-time data on radiation levels. This information is crucial for making informed decisions about evacuations, sheltering, and other protective actions.

Mobile Applications and Alerts: Developing mobile applications that deliver real-time alerts and safety information to citizens can enhance local preparedness. These apps can be integrated with national and international emergency systems, ensuring that the public receives timely and accurate information during a nuclear event.

Collaboration with International Agencies:

Accessing International Resources: Local governments can collaborate with international agencies like the IAEA and WHO to access resources, training, and expertise. This partnership can help enhance local capabilities and ensure that international safety protocols are effectively implemented at the community level.

Participating in International Exercises: Local authorities can participate in international nuclear emergency exercises, which simulate global responses to nuclear incidents. These exercises provide valuable experience and foster collaboration between different levels of government and international organizations.

Future Directions in International Fallout Safety

Advancements in Technology and Protocols:

Innovative Detection and Monitoring: Advances in technology, such as drones equipped with radiation sensors, AI-driven predictive modeling, and real-time global monitoring networks, are likely to enhance the effectiveness of international fallout safety protocols. These innovations will provide more accurate data and faster responses during nuclear events.

Updating Global Standards: As new technologies and research emerge, international organizations will need to update safety protocols and standards. Continuous improvement and adaptation of these guidelines will be essential to keep pace with the evolving landscape of nuclear safety.

Strengthening International Cooperation:

Global Partnerships: Strengthening international cooperation through global partnerships will be crucial in enhancing fallout safety protocols. Countries can work together to share resources, knowledge, and technology, creating a more unified and effective response to nuclear emergencies. This collaboration can also lead to the development of joint research initiatives, training programs, and mutual aid agreements, ensuring that all nations, regardless of their resources, can achieve a high level of preparedness.

Regional Safety Networks: In addition to global efforts, regional safety networks can play a significant role in improving fallout safety. Neighboring countries with shared geographic risks can collaborate to create regional emergency response plans, coordinate cross-border evacuations, and establish shared resources such as radiation detection equipment and medical facilities. These networks can also help standardize protocols across borders, reducing confusion and improving the overall effectiveness of the response.

Enhanced Information Sharing: Timely and transparent information sharing between countries and international organizations is essential for an effective global response to nuclear emergencies. This includes sharing real-time data on radiation levels, incident reports, and public health impacts. Enhancing global communication networks and developing platforms for rapid information exchange will help countries respond more effectively and minimize the impact of nuclear fallout on populations.

Preparing for Global Fallout Scenarios

Simulated Global Exercises: Conducting large-scale simulated global exercises can help countries and international organizations prepare for the complex challenges posed by nuclear fallout. These exercises can simulate a wide range of scenarios, from localized incidents to large-scale nuclear events with global implications. By participating in these exercises, countries can test their readiness, identify gaps in their response plans, and improve coordination with international partners.

Development of Global Emergency Stockpiles: Establishing global emergency stockpiles of essential resources, such as radiation protection gear, medical supplies, and food and water, can ensure that these items are available when needed, regardless of a country's resources. International organizations, such as the United Nations, could manage these stockpiles and distribute them quickly to affected regions during a nuclear emergency.

Promoting Public Awareness on a Global Scale: Increasing public awareness of nuclear fallout risks and safety protocols on a global scale is vital for ensuring that populations are prepared to respond effectively in an emergency. International organizations can lead global campaigns that educate the public about the dangers of nuclear fallout, the importance of following safety protocols, and the steps individuals can take to protect themselves and their families. These campaigns can be adapted to different cultural contexts and languages to reach a broad audience.

The Role of Technology in Future Fallout Safety Protocols

Advances in Predictive Analytics and AI: Predictive analytics and artificial intelligence (AI) can play a significant role in improving fallout safety protocols. These technologies can analyze vast amounts of data to predict the spread of nuclear fallout, identify populations at risk, and recommend optimal evacuation routes and sheltering strategies. AI can also assist in real-time decision-making during an emergency, helping authorities respond more quickly and effectively.

Integration of Drones and Robotics: Drones and robotics can be invaluable tools in assessing and responding to nuclear emergencies. Drones equipped with radiation sensors can quickly survey affected areas, providing real-time data on radiation levels and environmental conditions. Robotics can be used to perform tasks in highly radioactive environments, such as inspecting damaged infrastructure, repairing critical systems, or delivering supplies to isolated areas.

Enhanced Communication Technologies: The development of advanced communication technologies, such as satellite-based emergency communication systems, can ensure that information reaches even the most remote or affected areas during a nuclear event. These systems can be designed to withstand the impacts of a nuclear explosion, providing reliable communication channels for coordinating emergency responses and keeping the public informed.

Understanding and implementing international fallout safety protocols is essential for protecting populations and minimizing the impact of nuclear events. These protocols, developed and promoted by organizations such as the IAEA, WHO, and ICDO, provide a framework for global cooperation, information sharing, and emergency preparedness. By adopting these protocols and integrating them into national and local emergency plans, countries can enhance their ability to respond effectively to nuclear emergencies.

As technology advances and the global landscape of nuclear safety evolves, international protocols will need to be updated and strengthened. Future directions in fallout safety will likely involve greater international cooperation, the development of global and regional safety networks, and the integration of cutting-edge technologies such as AI, drones, and advanced communication systems.

By staying informed about international fallout safety protocols and preparing at both the national and local levels, governments, organizations, and individuals can contribute to a safer world where the risks of nuclear fallout are managed effectively, and populations are protected from the devastating effects of nuclear events.

Global Perspectives on Nuclear Fallout Preparedness

Nuclear fallout preparedness varies significantly across the globe, influenced by a nation's history, geographical location, economic resources, and political landscape. Understanding these global perspectives provides valuable insights into how different countries approach the challenge of preparing for nuclear events. This chapter explores the diverse approaches to nuclear fallout preparedness, highlights key strategies employed by various nations, and examines the cultural, political, and technological factors that shape these strategies.

Historical Context and Its Influence on Preparedness

The Cold War Legacy:

United States and Russia: The Cold War era left a lasting impact on nuclear fallout preparedness in both the United States and Russia (formerly the Soviet Union). During this period, both nations developed extensive civil defense programs, including public fallout shelters, mass evacuation plans, and widespread public education on nuclear safety. Although the intensity of these programs has diminished since the end of the Cold War, the infrastructure and cultural awareness established during this time continue to influence current preparedness efforts.

Eastern Europe: Many Eastern European countries, formerly part of the Soviet bloc, also maintain a legacy of Cold War-era civil defense measures. Nations like Poland, Hungary, and the Czech Republic have retained some elements of their old civil defense programs, though these have often been scaled down or repurposed in the post-Cold War era.

The Impact of Nuclear Incidents:

Japan and the Fukushima Disaster: Japan's approach to nuclear fallout preparedness is heavily influenced by the Fukushima Daiichi nuclear disaster in 2011. This event highlighted the vulnerabilities in Japan's nuclear safety protocols and led to significant reforms, including stricter regulations, improved early warning systems, and increased public education on radiation safety. Japan has also focused on enhancing evacuation plans and sheltering strategies, particularly for communities near nuclear power plants.

Chernobyl and Ukraine: The Chernobyl disaster of 1986 had a profound impact on Ukraine and neighboring countries. It exposed the dangers of inadequate safety measures and poor emergency response planning. In response, Ukraine and other affected countries have since worked to improve their nuclear safety standards, develop better monitoring systems, and educate the public on the risks of radiation exposure.

Geopolitical and Economic Factors

Nuclear Powers and Their Preparedness:

United States: As a major nuclear power, the United States has developed a comprehensive approach to nuclear fallout preparedness. This includes the Federal Emergency Management Agency (FEMA) coordinating national response plans, the Department of Homeland Security (DHS) overseeing preparedness programs, and the Department of Energy (DOE) managing nuclear safety. Public education campaigns, drills, and the development of advanced technology for radiation detection and protection are central to U.S. preparedness efforts.

Russia: Russia maintains a robust civil defense system, with the Ministry of Emergency Situations (EMERCOM) playing a key role in disaster preparedness and response. Russia's approach emphasizes large-scale evacuation plans, public shelters, and the use of military resources in the event of a nuclear emergency. Given Russia's vast territory and nuclear arsenal, the country prioritizes maintaining a high level of preparedness across its regions.

Non-Nuclear States:

Switzerland: Switzerland is renowned for its comprehensive civil defense program, despite not being a nuclear power. The country has invested heavily in public fallout shelters, with enough capacity to protect its entire population. Swiss law mandates that all new buildings include shelter space, and the government regularly updates its preparedness plans and public education initiatives.

Sweden: Sweden also has a strong tradition of civil defense, with extensive fallout shelter networks and public education programs. Sweden's preparedness is rooted in its policy of neutrality and self-reliance, aiming to protect its population in the event of a nuclear conflict or accident, despite not possessing nuclear weapons.

Developing Countries:

India and Pakistan: As nuclear-armed neighbors with a history of conflict, India and Pakistan both face unique challenges in nuclear fallout preparedness. Both countries have made efforts to develop civil defense measures, including public awareness campaigns and the establishment of emergency response agencies. However, economic constraints and the need to address other pressing issues, such as poverty and infrastructure development, limit the extent of their preparedness programs.

Sub-Saharan Africa: In much of Sub-Saharan Africa, nuclear fallout preparedness is less developed due to limited resources, low risk perception, and other pressing public health and safety concerns. However, international organizations and NGOs have worked to improve radiation monitoring and emergency response capabilities, particularly in countries with uranium mining activities or nuclear research facilities.

Cultural and Societal Influences

Public Perception and Preparedness:

Asia: In countries like Japan and South Korea, public perception of nuclear risks is shaped by both historical experiences (such as Hiroshima, Nagasaki, and Fukushima) and ongoing geopolitical tensions. As a result, these nations place a high emphasis on public education, regular drills, and the availability of personal protective equipment. The public in these countries tends to be more aware of nuclear risks and more willing to participate in preparedness activities.

Western Europe: In Western Europe, public perception of nuclear fallout varies. Countries like Germany and Austria, which have strong anti-nuclear movements, tend to focus on phasing out nuclear power and investing in renewable energy. These countries also emphasize the importance of international cooperation in nuclear safety, but public preparedness for fallout is often less emphasized compared to countries with a history of civil defense programs.

Education and Information Dissemination:

Scandinavian Countries: Countries like Norway and Finland emphasize public education on nuclear safety as part of their broader civil defense strategies. Information on radiation risks, protective measures, and emergency procedures is integrated into school curricula and public campaigns. The focus on transparency and public trust helps ensure that citizens are well-informed and prepared to respond to nuclear emergencies.

Latin America: In Latin American countries, nuclear fallout preparedness is generally less prominent in public discourse, partly due to the lower perceived risk of nuclear conflict or accidents. However, nations with nuclear energy programs, like Brazil and Argentina, have made efforts to educate the public about radiation safety and emergency response, often in collaboration with international agencies.

Technological and Infrastructure Considerations

Advanced Technology in Preparedness:

United States and Japan: Both the United States and Japan are leaders in the development and deployment of advanced technology for nuclear fallout preparedness. This includes sophisticated radiation detection networks, automated early warning systems, and AI-driven predictive models for fallout dispersion. These technologies are integrated into national emergency management frameworks, providing real-time data to guide decision-making during a nuclear event.

European Union: The European Union (EU) promotes the use of technology in nuclear safety through its Joint Research Centre (JRC) and other initiatives. The EU has invested in cross-border radiation monitoring systems, shared databases, and collaborative research on nuclear safety technologies. Member states benefit from these shared resources, enhancing their overall preparedness.

Infrastructure and Resource Allocation:

Australia and New Zealand: Australia and New Zealand, while geographically distant from major nuclear powers, have nonetheless invested in infrastructure to protect against nuclear fallout. This includes public shelters, emergency response facilities, and stockpiles of essential supplies. Both countries also emphasize the importance of resilient infrastructure, such as robust power grids and communication networks, to maintain functionality during a nuclear event.

China: China has made significant investments in civil defense infrastructure, including the construction of urban fallout shelters and the development of emergency evacuation plans for densely populated areas. China's approach combines traditional civil defense measures with modern technology, such as mobile apps for emergency alerts and digital platforms for public education.

International Cooperation and Knowledge Sharing

Regional Collaborations:

NATO and EU: NATO and the EU both play significant roles in facilitating regional cooperation on nuclear fallout preparedness. NATO's Civil Emergency Planning Committee (CEPC) coordinates efforts among member states to develop and share best practices, conduct joint exercises, and standardize emergency response protocols.

The EU's mechanisms, such as the European Civil Protection Mechanism, support coordinated responses to nuclear incidents across member states.

ASEAN: The Association of Southeast Asian Nations (ASEAN) has also focused on improving regional cooperation in nuclear safety. Through the ASEAN Network of Regulatory Bodies on Atomic Energy (ASEANTOM), member states collaborate on developing regulatory frameworks, conducting joint training programs, and sharing information on nuclear safety.

Global Partnerships:

International Atomic Energy Agency (IAEA): The IAEA remains the central organization for global cooperation on nuclear safety. It provides technical assistance, conducts peer reviews, and facilitates the exchange of knowledge and best practices among member states. The IAEA's Incident and Emergency Centre (IEC) plays a crucial role in coordinating international responses to nuclear emergencies.

United Nations and WHO: The United Nations and the World Health Organization (WHO) also contribute to global nuclear fallout preparedness through initiatives that promote public health, environmental monitoring, and emergency response. These organizations work with member states to implement international safety standards and improve global resilience to nuclear threats.

Future Directions in Global Nuclear Fallout Preparedness

Adapting to New Threats:

Cybersecurity Concerns: As nuclear facilities and emergency management systems become more reliant on digital technology, cybersecurity has emerged as a critical aspect of nuclear fallout preparedness. Countries must invest in protecting their nuclear infrastructure from cyberattacks that could disrupt safety systems or emergency responses.

Climate Change and Natural Disasters: The intersection of nuclear safety and climate change is another emerging concern. Rising sea levels, increased storm intensity, and other climate-related risks could impact nuclear facilities and complicate emergency responses. Nations are increasingly factoring these risks into their nuclear preparedness strategies.

Advancing Public Education:

Global Education Campaigns: International organizations and national governments are recognizing the need for ongoing public education campaigns that address the evolving nature of nuclear threats. These campaigns aim to improve public understanding of radiation risks, promote protective behaviors, and **empower communities** to take proactive steps in preparing for nuclear fallout. By leveraging modern communication platforms, including social media, mobile apps, and online learning, these campaigns can reach a wider audience and provide accessible, up-to-date information on how to respond effectively in a nuclear emergency.

Incorporating Nuclear Preparedness into School Curricula: To ensure that future generations are better prepared, some countries are integrating nuclear safety and fallout preparedness into their school curricula. This education begins at an early age, teaching students about the basics of radiation, how to protect themselves during a nuclear event, and the importance of emergency preparedness. These programs help foster a culture of safety and preparedness that can be passed on to families and communities.

Community-Driven Initiatives: In addition to government-led efforts, there is a growing recognition of the role that communities can play in nuclear fallout preparedness. Grassroots initiatives, such as neighborhood preparedness groups, local training sessions, and volunteer emergency response teams, are becoming more common. These initiatives empower individuals to take ownership of their safety and contribute to the resilience of their communities.

Global Equity in Fallout Preparedness

Addressing Disparities in Resources:

Developing Countries: One of the most significant challenges in global nuclear fallout preparedness is the disparity in resources between developed and developing countries. While wealthier nations can invest in advanced technologies, infrastructure, and public education, many developing countries struggle to allocate resources to nuclear safety. Addressing these disparities requires international assistance, capacity-building programs, and the sharing of technology and expertise.

International Aid and Support:

Global Funds for Preparedness: Establishing global funds dedicated to nuclear fallout preparedness can help bridge the gap between nations with different resource levels. These funds could support the construction of shelters, the distribution of protective equipment, and the development of emergency plans in countries that lack the financial means to do so independently.

Knowledge and Technology Transfer: Wealthier nations and international organizations can assist developing countries by providing access to radiation detection equipment, medical supplies, and training programs. This transfer of knowledge and technology can significantly enhance the ability of developing nations to prepare for and respond to nuclear fallout.

Promoting Inclusive Preparedness Strategies:

Culturally Sensitive Approaches: Ensuring that nuclear fallout preparedness strategies are inclusive and culturally sensitive is critical for their effectiveness. Preparedness plans should consider the specific needs of vulnerable populations, including children, the elderly, and those with disabilities. Tailoring public education campaigns and emergency plans to accommodate cultural differences and local languages can improve community engagement and ensure that all members of society are adequately prepared.

Global Solidarity and Cooperation:

Building International Solidarity: In the face of global nuclear threats, building a sense of international solidarity is essential. Countries must recognize that nuclear fallout is a transnational issue that requires collective action. By fostering a spirit of cooperation and mutual support, the global community can work together to ensure that all nations, regardless of their economic status, are equipped to protect their populations from the dangers of nuclear fallout.

Case Studies in Global Nuclear Fallout Preparedness

Japan's Comprehensive Approach Post-Fukushima:

Reforms and Innovations: After the Fukushima disaster, Japan undertook significant reforms to enhance its nuclear safety and fallout preparedness. These efforts included the introduction of stricter regulatory standards, the development of advanced early warning systems, and a renewed focus on public education and community engagement. Japan's experience provides valuable lessons for other countries seeking to improve their preparedness in the wake of a nuclear incident.

Switzerland's Extensive Shelter Network:

A Model for Public Shelters: Switzerland's approach to fallout preparedness, characterized by its extensive network of public shelters, serves as a model for other nations. The Swiss government's commitment to maintaining a shelter capacity that can accommodate its entire population, along with regular public education campaigns and drills, highlights the importance of infrastructure and public engagement in effective preparedness strategies.

South Korea's Integrated Civil Defense System:

Balancing Modern Threats with Traditional Measures: South Korea has developed a sophisticated civil defense system that integrates modern technology with traditional civil defense measures. The country's approach includes advanced early warning systems, comprehensive public education programs, and a network of shelters and evacuation routes. South Korea's preparedness efforts are driven by its unique geopolitical situation, underscoring the importance of tailoring strategies to local conditions.

Finland's Resilient Infrastructure and Public Engagement:

Preparedness in the Nordic Context: Finland's approach to nuclear fallout preparedness emphasizes resilient infrastructure, public trust, and widespread community involvement. The country's focus on building robust shelters, integrating preparedness into the education system, and engaging the public through transparent communication has made it one of the most prepared nations in Europe.

The Future of Global Nuclear Fallout Preparedness

Innovations in Technology and Policy:

Emerging Technologies: The future of nuclear fallout preparedness will likely be shaped by advancements in technology, including AI-driven predictive models, drone-based radiation detection, and real-time communication platforms. These technologies will enhance the ability of governments and communities to respond quickly and effectively to nuclear threats.

Global Policy Developments: International policy developments, such as new treaties, conventions, and agreements on nuclear safety, will also play a critical role in shaping the future of fallout preparedness. These policies can promote greater cooperation, standardize safety protocols, and ensure that all nations are working towards common goals in nuclear safety.

Resilience in the Face of New Challenges:

Climate Change and Nuclear Safety: As climate change presents new challenges, such as extreme weather events and rising sea levels, countries will need to adapt their nuclear fallout preparedness strategies. This may involve reassessing the location of nuclear facilities, updating infrastructure to withstand natural disasters, and integrating climate risk assessments into emergency planning.

Strengthening International Cooperation: The global nature of nuclear threats requires a renewed commitment to international cooperation. Strengthening global networks, improving information sharing, and fostering mutual support will be essential for building resilience in the face of future nuclear challenges.

Global perspectives on nuclear fallout preparedness reflect the diverse approaches, challenges, and innovations that countries around the world have developed in response to the threat of nuclear events. From the legacy of the Cold War to the lessons learned from recent nuclear incidents, nations have crafted unique strategies based on their historical experiences, geopolitical realities, and available resources.

As the world continues to face evolving nuclear threats, the importance of international cooperation, technological innovation, and public engagement in nuclear fallout preparedness cannot be overstated. By learning from each other's experiences and working together to address common challenges, the global community can enhance its collective ability to protect populations and mitigate the risks associated with nuclear fallout.

The future of global nuclear fallout preparedness will be shaped by our ability to adapt to new threats, embrace emerging technologies, and promote equity in preparedness efforts. Through continued collaboration and a shared commitment to safety, the world can build a more resilient and secure future for all.

Fallout Shelter Maintenance: Keeping It Functional

Maintaining a fallout shelter is essential to ensure that it remains functional and ready for use in the event of a nuclear emergency. Regular maintenance not only preserves the shelter's structural integrity but also ensures that all systems, supplies, and equipment are in working order. This chapter will provide a comprehensive guide to fallout shelter maintenance, covering key aspects such as structural inspections, system checks, supply management, and long-term upkeep strategies.

Structural Inspections

Routine Structural Assessments:

Wall and Roof Integrity: Regularly inspect the walls and roof of your shelter for any signs of damage, such as cracks, leaks, or water infiltration. These issues can compromise the shelter's ability to protect against radiation and environmental hazards. Address any structural damage immediately to prevent further deterioration.

Foundation Stability: Check the foundation for signs of settling, cracks, or moisture. A stable foundation is critical for the long-term durability of your shelter. If you notice any significant changes, consult a structural engineer to assess the situation and recommend necessary repairs.

Moisture and Waterproofing:

Waterproofing Systems: Ensure that your shelter's waterproofing systems, such as drainage, sealants, and membranes, are functioning properly. Water infiltration can lead to mold growth, weaken the structure, and damage stored supplies. Regularly check for leaks and reseal any areas where water might penetrate.

Humidity Control: Maintain proper humidity levels inside the shelter to prevent condensation, which can lead to rust, mold, and material degradation. Consider using dehumidifiers or moisture-absorbing materials like silica gel to control humidity.

Ventilation and Air Quality:

Ventilation System Checks: Regularly inspect and clean the ventilation system to ensure that it is free of blockages and functioning efficiently. Proper ventilation is essential for maintaining air quality and removing contaminants from the shelter. Replace air filters as needed and ensure that all fans, ducts, and vents are in good condition.

Air Quality Monitoring: Use air quality monitors to check for the presence of harmful gases, such as carbon monoxide or radon. These devices can alert you to potential hazards that may require immediate attention. Ensure that all detection equipment is calibrated and functioning correctly.

System Maintenance

Power Supply Systems:

Generator Maintenance: If your shelter relies on a generator for power, perform regular maintenance checks, including oil changes, fuel filter replacements, and battery inspections. Test the generator periodically to ensure that it starts and runs smoothly. Keep a supply of fuel on hand, and use fuel stabilizers to prevent degradation.

Battery Backup Systems: Inspect battery backup systems to ensure they are fully charged and operational. Replace any batteries that show signs of wear or reduced capacity. Regularly test the system to verify that it can provide power in the event of a generator failure.

Water and Sanitation Systems:

Water Storage and Filtration: Check your water storage containers for leaks or signs of contamination. Rotate stored water periodically to ensure freshness. Inspect and maintain water filtration systems, replacing filters as recommended by the manufacturer.

Sanitation Equipment: Inspect portable toilets, waste disposal systems, and plumbing fixtures for functionality and cleanliness. Regularly empty waste containers, and sanitize all equipment to prevent the buildup of harmful bacteria. Ensure that your shelter has an adequate supply of sanitation chemicals and cleaning materials.

Heating, Cooling, and Ventilation:

Climate Control Systems: Test heating, cooling, and ventilation systems to ensure they can maintain a comfortable temperature inside the shelter. Clean or replace air filters in HVAC units, and check for any blockages or mechanical issues in the system. If your shelter uses a wood stove or other heat source, inspect it for safety and efficiency.

Emergency Ventilation: Make sure that any manual or backup ventilation systems, such as hand-crank air pumps, are in good working order. These systems are essential if your primary ventilation fails.

Supply Management

Food and Water Supplies:

Food Rotation and Inventory: Regularly check your food supplies for expiration dates and rotate them to ensure that all items remain within their shelf life. Maintain an up-to-date inventory of all food items, including quantities and expiration dates, and replace any items that are nearing their expiration.

Water Reserves: Ensure that your water reserves are sufficient for the expected duration of shelter use. Replace stored water every six months, and consider using water treatment tablets or filters to extend the shelf life of stored water.

Medical Supplies:

First Aid Kit Maintenance: Inspect your first aid kit and medical supplies to ensure that they are complete and in good condition. Replace any expired medications, bandages, or other perishable items. Consider stocking additional medical supplies, such as antibiotics, pain relievers, and antiseptics, to address potential health issues during extended shelter stays.

Specialty Medical Equipment: If your shelter contains specialty medical equipment, such as defibrillators or oxygen tanks, regularly check that they are functional and fully charged or filled. Ensure that all equipment is stored correctly and is easily accessible in an emergency.

Emergency Tools and Equipment:

Tool Inspection: Inspect all emergency tools and equipment, such as flashlights, radios, fire extinguishers, and hand tools, to ensure they are in good working order. Replace batteries in electronic devices regularly, and keep a supply of spare batteries on hand.

Communication Devices: Test communication devices, including two-way radios, satellite phones, and emergency broadcast radios, to ensure they are operational. Maintain backup communication methods in case primary systems fail.

Long-Term Upkeep Strategies

Regular Drills and Training:

Emergency Drills: Conduct regular emergency drills to ensure that all shelter occupants know how to use the equipment, access supplies, and respond to various emergency scenarios. Drills should simulate realistic conditions and include practice in using backup systems, such as manual ventilation or alternative power sources.

Skill Maintenance: Keep your skills sharp by regularly practicing essential tasks, such as first aid, fire suppression, and equipment repairs. Consider taking refresher courses or workshops to stay updated on best practices in emergency preparedness.

Documentation and Record Keeping:

Maintenance Log: Keep a detailed maintenance log that records all inspections, repairs, and system tests. This log should include dates, findings, and any actions taken. Regularly reviewing this log can help identify patterns and potential issues before they become critical.

Supply Inventory: Maintain a current inventory of all supplies, including food, water, medical items, tools, and equipment. Regularly update this inventory as items are used, replaced, or added. An accurate inventory ensures that you can quickly assess your shelter's readiness in an emergency.

Shelter Upgrades and Improvements:

Periodic Assessments: Periodically assess the overall condition of your shelter and identify areas where improvements can be made. This might include upgrading ventilation systems, reinforcing structural elements, or adding new technology to enhance the shelter's capabilities.

Budgeting for Upgrades: Set aside a portion of your budget for ongoing maintenance and potential upgrades. Prioritize critical areas, such as structural integrity and essential systems, but also consider investing in comfort and sustainability improvements, such as better insulation or renewable energy sources.

Preparing for Seasonal Changes

Winterization:

Insulation Checks: Before winter, inspect the insulation in your shelter to ensure it is adequate for maintaining warmth. Add extra insulation to walls, ceilings, and doors if needed. Also, check for drafts around doors and windows and seal any gaps to prevent heat loss.

Heating System Maintenance: Test and service your heating systems, such as stoves, heaters, or heat pumps, to ensure they are ready for use during cold weather. Make sure you have enough fuel or power for extended use.

Summer Readiness:

Cooling System Maintenance: Ensure that cooling systems, such as fans or air conditioning units, are working efficiently. Clean or replace filters, check refrigerant levels, and test the system's ability to maintain a comfortable temperature in hot weather.

Ventilation Adjustments: Adjust ventilation systems to optimize airflow and prevent overheating. In hot climates, consider adding shading or reflective materials to reduce heat gain inside the shelter.

Addressing Common Maintenance Challenges

Pest Control:

Preventive Measures: Implement preventive measures to keep pests out of your shelter, such as sealing cracks and gaps, using pest repellent, and storing food in airtight containers. Regularly inspect the shelter for signs of pests, such as droppings or damage to supplies.

Pest Management: If pests are detected, take immediate action to remove them and prevent further infestation. This may involve setting traps, using pest control products, or seeking professional help if the problem is severe.

Mold and Mildew Prevention:

Moisture Control: Maintain proper moisture control through effective ventilation, humidity management, and waterproofing. Use mold-resistant materials in the construction and finishing of your shelter, and regularly inspect for signs of mold or mildew.

Cleaning and Remediation: If mold or mildew is detected, clean the affected area immediately using mold-killing products. Remove and replace any materials that are heavily infested. Address the underlying moisture issue to prevent future growth.

Aging Infrastructure:

Proactive Maintenance: As your shelter ages, more frequent inspections and proactive maintenance will be necessary to keep it in good condition. Regularly update older systems, materials, and equipment to ensure they remain functional and safe.

Professional Inspections: Consider having a professional conduct periodic inspections, particularly if your shelter is older or if you are concerned about specific issues like structural integrity or electrical safety. A professional assessment can identify problems that might not be apparent during routine checks.

Maintaining a fallout shelter is a critical responsibility that ensures the shelter remains functional, safe, and ready for use in the event of a nuclear emergency. Regular maintenance activities, including structural inspections, system checks, and supply management, are essential for preserving the integrity and operational readiness of the shelter. By implementing long-term upkeep strategies and addressing common maintenance challenges, you can extend the lifespan of your shelter and ensure that it provides adequate protection when needed.

Final Considerations

Commitment to Ongoing Maintenance:

Routine and Consistency: Maintenance is not a one-time task but an ongoing process that requires routine attention. Set a regular schedule for inspections and system tests to keep the shelter in top condition. Consistency in these efforts will prevent minor issues from becoming major problems.

Involving All Occupants: Ensure that all shelter occupants are familiar with the maintenance routines and know how to perform basic checks and repairs. This shared responsibility increases the overall readiness of the shelter and ensures that everyone is prepared to contribute in an emergency.

Preparing for the Unexpected:

Emergency Repairs: Keep a well-stocked toolkit and spare parts on hand for emergency repairs. Being prepared to address unexpected issues, such as a burst pipe or a malfunctioning generator, is critical for maintaining the shelter's functionality during a crisis.

Adaptability: Be prepared to adapt your maintenance strategies as new challenges arise. Whether it's due to aging infrastructure, changing climate conditions, or new technological developments, flexibility in your approach will help ensure that your shelter remains effective over time.

Documentation and Knowledge Transfer:

Detailed Records: Maintain detailed records of all maintenance activities, upgrades, and repairs. This documentation will be invaluable for tracking the shelter's condition and planning future maintenance tasks.

Knowledge Transfer: Ensure that knowledge of the shelter's maintenance routines is passed on to future occupants or family members. This transfer of knowledge ensures that the shelter remains functional and that its maintenance routines are upheld, even if there is a change in who is responsible for its upkeep.

Continual Improvement:

Embracing Innovation: As new materials, technologies, and best practices emerge, consider incorporating them into your shelter's design and maintenance plan. Staying informed about advancements in shelter construction and emergency preparedness can help you make informed decisions about upgrades and improvements.

Feedback and Assessment: After each maintenance cycle, assess what worked well and what could be improved. Gathering feedback from all shelter occupants can help refine your maintenance routines and ensure that the shelter meets everyone's needs effectively.

Maintaining a fallout shelter is a proactive and ongoing commitment that plays a crucial role in ensuring the safety and well-being of its occupants. By diligently performing regular inspections, system checks, and supply management, and by addressing maintenance challenges as they arise, you can keep your shelter in optimal condition. Through careful planning, consistent upkeep, and a willingness to adapt and improve, your fallout shelter will remain a reliable and secure refuge in the event of a nuclear emergency.

Re-entry into the World: When Is It Safe?

Re-entry into the world after a nuclear event is a critical and complex decision that requires careful consideration of various factors, including radiation levels, environmental conditions, and the availability of resources. Knowing when it is safe to leave the shelter and how to do so effectively is vital for ensuring the health and safety of all shelter occupants. This chapter will guide you through the process of determining when it is safe to re-enter the outside world, what precautions to take, and how to navigate the challenges that may arise during this transition.

Understanding Radiation Decay and Safe Levels

Radiation Decay Process:

Initial Fallout: After a nuclear detonation, the environment outside the shelter will be heavily contaminated with radioactive particles. These particles emit radiation, which poses significant health risks. The intensity of radiation decreases over time, but the initial days and weeks after the event are the most dangerous.

The Rule of Sevens: The "7-10 Rule" is a basic principle used to estimate radiation decay. According to this rule, for every sevenfold increase in time after the explosion, the radiation level decreases by a factor of ten. For example, if the radiation level is 1,000 roentgens per hour immediately after the explosion, it will drop to 100 roentgens per hour after 7 hours, and 10 roentgens per hour after 49 hours. However, even at lower levels, prolonged exposure can still be harmful.

Safe Radiation Levels for Re-entry:

Threshold Levels: The general guideline for safe re-entry into the environment is when radiation levels fall below 0.5 roentgens per hour (500 milliroentgens per hour). At this level, the risk of acute radiation sickness is minimal, and with proper precautions, it is safe to spend limited time outside the shelter.

Long-Term Exposure Considerations: While lower radiation levels may allow for temporary re-entry, long-term exposure to even small amounts of radiation can increase the risk of cancer and other health issues. It's crucial to monitor radiation levels continuously and limit time spent outside until levels are deemed safe for permanent habitation.

Tools for Measuring Radiation

Radiation Detection Equipment:

Geiger Counters: Geiger counters are the most common tools used to measure radiation levels in the environment. They detect and measure ionizing radiation, providing real-time data on the level of contamination. Ensure that your Geiger counter is calibrated and in good working order before using it to assess the safety of re-entry.

Dosimeters: Personal dosimeters measure cumulative radiation exposure over time. These devices are worn by individuals and provide an ongoing record of the amount of radiation they have been exposed to. This is particularly important for monitoring long-term exposure as you begin re-entry.

Survey Meters: Survey meters are more advanced radiation detection devices that can provide more precise measurements of different types of radiation, including alpha, beta, gamma, and neutron radiation. They are useful for assessing the specific hazards in the environment.

Using Detection Equipment:

Establishing a Baseline: Before leaving the shelter, take baseline readings of radiation levels both inside and immediately outside the shelter. This will help you compare changes over time and determine if conditions are improving.

Monitoring Progress: As you begin to explore the area outside the shelter, use your detection equipment to monitor radiation levels continuously. Pay particular attention to areas where radiation might be more concentrated, such as low-lying areas, water sources, or places where fallout might have accumulated.

Preparing for Re-entry

Personal Protective Equipment (PPE):

Radiation-Proof Clothing: Before exiting the shelter, ensure that all occupants are equipped with radiation-proof clothing, such as lead-lined or specially designed protective suits. These garments help minimize exposure to radioactive particles and should be worn whenever outside the shelter until radiation levels are confirmed to be safe.

Respiratory Protection: Wear respirators or masks with high-efficiency particulate air (HEPA) filters to protect against inhaling radioactive dust and particles. Ensure that these masks are properly fitted and that the filters are in good condition.

Decontamination Procedures:

Pre-Exit Decontamination: Before leaving the shelter, wash thoroughly to remove any potential radioactive particles that may have entered the shelter. This reduces the risk of contaminating your protective gear or increasing exposure while outside.

Post-Re-entry Decontamination: Upon returning to the shelter, remove protective clothing carefully and follow decontamination procedures. This includes washing the skin with soap and water, safely disposing of contaminated gear, and monitoring yourself and others for any signs of radiation exposure.

Communication and Navigation:

Establishing Communication: Before re-entry, ensure that you have reliable communication devices, such as radios or satellite phones, to stay in contact with other shelter occupants or external emergency services. This is crucial for coordinating movements, receiving updates on radiation levels, and reporting any issues.

Mapping and Planning: Create a map of the area around your shelter and plan your movements carefully. Avoid areas known to have high radiation levels or those that are difficult to navigate. Identify potential safe zones or locations where you can find additional shelter or resources.

Assessing Environmental Conditions

Water and Food Safety:

Water Sources: Test local water sources for radiation contamination before using them for drinking, cooking, or cleaning. Water contaminated with radioactive particles can pose serious health risks. Use water filtration systems designed to remove radioactive contaminants, or rely on stored water until you are certain that natural sources are safe.

Food Safety: Avoid consuming food that may have been exposed to radioactive fallout. This includes crops, wild plants, and animals. If you must rely on local food sources, test them for contamination using a Geiger counter or other detection devices, and prioritize foods that have been stored or preserved in sealed containers.

Structural Integrity of Buildings:

Assessing Damage: As you begin to explore outside the shelter, assess the structural integrity of buildings and infrastructure. Many structures may have been damaged by the initial blast, shockwave, or radiation, making them unsafe to enter. Avoid entering buildings that show signs of severe damage, such as cracks, leaning walls, or collapsed roofs.

Identifying Safe Structures: Look for buildings that appear structurally sound and free from visible damage. These structures can provide temporary shelter or serve as staging areas as you transition back to the outside world. However, continue to monitor radiation levels inside these buildings, as they may still harbor radioactive contamination.

Environmental Hazards:

Chemical and Biological Hazards: In addition to radiation, be aware of other potential hazards in the environment, such as chemical spills, fires, or biological contamination. The disruption caused by a nuclear event can lead to secondary hazards that pose additional risks to health and safety.

Wildlife and Natural Hazards: Be cautious of wildlife that may have been affected by radiation or changes in the environment. Animals may behave unpredictably, and some species may be more dangerous due to increased aggression or illness. Additionally, natural hazards such as landslides, flooding, or fallen debris may pose risks as you move through the environment.

Psychological and Social Considerations

Managing Anxiety and Stress:

Mental Health Awareness: Re-entering the world after a nuclear event can be a stressful and anxiety-inducing experience. It's important to be aware of the mental health impacts on yourself and others. Provide support and encouragement, and take breaks if anyone is feeling overwhelmed.

Group Cohesion: Maintain strong communication within your group to ensure that everyone feels supported and informed. Decisions should be made collectively, taking into account the physical and emotional well-being of all shelter occupants.

Social Dynamics and Safety:

Interacting with Others: If you encounter other survivors, approach with caution. In the aftermath of a nuclear event, resources may be scarce, and some individuals or groups may react unpredictably. Establish trust carefully, and be prepared to defend yourself if necessary.

Community Building: As you re-enter the world, consider the importance of rebuilding a community. Working together with other survivors can enhance safety, provide mutual support, and improve the chances of long-term survival. Sharing resources, knowledge, and skills can help create a more stable and resilient post-event society.

Long-Term Re-entry Strategies

Gradual Re-entry:

Short-Term Excursions: Initially, limit your time outside the shelter to short excursions for essential tasks, such as gathering supplies, assessing conditions, and exploring safe zones. This reduces the risk of prolonged radiation exposure and allows you to monitor changes in environmental conditions.

Progressive Exploration: As radiation levels continue to decrease, gradually increase the duration and range of your excursions. This phased approach allows you to adapt to the outside environment while minimizing risks.

Permanent Re-entry:

Assessing Habitability: Before committing to permanent re-entry, conduct a thorough assessment of the environment to ensure it is safe for long-term habitation. This includes verifying that radiation levels are consistently low, that water and food sources are safe, and that shelter and infrastructure are secure.

Establishing a New Normal: Rebuilding life after a nuclear event requires creating new routines, adapting to changed conditions, and prioritizing safety. Focus on establishing stable sources of food, water, and shelter, and gradually work toward rebuilding community and infrastructure.

Rebuilding and Recovery:

Resource Management: Carefully manage any resources you acquire during re-entry. Prioritize sustainable practices, such as rationing food, purifying water, and conserving energy, to ensure long-term survival.

Planning for the Future: As you settle into a post-nuclear environment, begin planning for long-term recovery and rebuilding. This includes considering how to restore infrastructure, re-establish agriculture, and create a self-sustaining community. Recovery will be a gradual process that requires careful planning, resource management, and collaboration with others who have survived the event.

Establishing a Sustainable Post-Nuclear Life

Agriculture and Food Production:

Assessing Soil Safety: Before attempting to grow crops, test the soil for radiation contamination. Some types of radiation, such as cesium-137, can remain in the soil for years, affecting the safety of food grown in it. If the soil is contaminated, consider using raised beds with clean soil or hydroponic systems to grow food safely.

Selecting Crops: Choose crops that are known for their resilience and fast growth. Root vegetables, for example, may be more resistant to radiation, while leafy greens can be grown quickly in controlled environments. Focus on crops that provide essential nutrients and calories.

Water Management:

Ensuring Clean Water: A reliable source of clean water is essential for long-term survival. Continue to test water sources regularly for contamination, and use filtration and purification methods to ensure safety. Rainwater collection systems can also provide a renewable source of water, though it's important to ensure that rainwater is free from airborne radioactive particles.

Irrigation and Conservation: Develop irrigation systems that make efficient use of available water. Drip irrigation, for example, minimizes water waste and can be used to sustain crops in a post-nuclear environment. Additionally, practice water conservation methods to ensure that your supply lasts as long as possible.

Shelter and Infrastructure Development:

Building Materials: When constructing new shelters or repairing existing ones, choose building materials that offer radiation protection and structural integrity. Materials like concrete, brick, and earth provide good shielding from residual radiation.

Renewable Energy: As traditional power sources may be unreliable, invest in renewable energy options such as solar panels or wind turbines. These systems can provide the electricity needed for lighting, water purification, and communication devices, reducing your reliance on potentially scarce fuel supplies.

Health and Medical Care:

Monitoring Health: Continue to monitor the health of all group members for signs of radiation sickness, malnutrition, or other health issues that could arise in a post-nuclear environment. Early detection and treatment are crucial for maintaining the health of the community.

Medical Supplies: Maintain a well-stocked medical kit and seek out additional supplies when possible. Focus on acquiring antibiotics, pain relievers, and treatments for radiation exposure. If possible, establish a relationship with a medical professional who can provide guidance and care.

Community and Social Structure:

Building Community: In the aftermath of a nuclear event, rebuilding a sense of community is vital for emotional support and mutual aid. Establish clear communication and decision-making structures within your group, and work together to set goals and priorities for survival and recovery.

Leadership and Cooperation: Effective leadership and cooperation are essential for navigating the challenges of a post-nuclear world. Encourage collaboration, distribute responsibilities fairly, and foster an environment of trust and shared purpose.

Planning for Future Risks

Continual Monitoring:

Ongoing Radiation Monitoring: Even after initial re-entry, continue to monitor radiation levels regularly. This ensures that you can detect any changes in environmental conditions that might pose a new threat. Use a combination of fixed monitoring stations and portable devices to keep track of radiation levels in different areas.

Environmental Changes: Stay vigilant for other environmental changes, such as shifts in weather patterns, the emergence of new hazards, or the impact of wildlife on your settlement. Being aware of these changes allows you to adapt your strategies and protect your community.

Preparing for Secondary Threats:

Potential Secondary Events: Be prepared for secondary threats that could arise after a nuclear event, such as chemical spills, natural disasters, or conflicts over resources. Develop contingency plans for these scenarios and ensure that your group knows how to respond.

Emergency Drills: Continue to conduct regular emergency drills to keep everyone in the community prepared for potential threats. These drills should cover a range of scenarios, including evacuation, sheltering in place, and responding to medical emergencies.

Long-Term Resilience:

Diversifying Resources: To build long-term resilience, focus on diversifying your resources. This includes growing a variety of crops, securing multiple water sources, and developing alternative energy options. A diverse resource base reduces the risk of dependency on any single supply, making your community more adaptable to future challenges.

Education and Skills Development: Invest in education and skills development for all community members. Knowledge of first aid, agriculture, engineering, and leadership will be crucial for surviving and thriving in a post-nuclear world. Encourage continuous learning and the sharing of knowledge within the group.

Re-entering the world after a nuclear event is a complex process that requires careful planning, preparation, and ongoing vigilance. As you rebuild and adapt to the new realities of a post-nuclear world, focus on sustainability, resource management, and community resilience.

Rebuilding Society after a Nuclear Event

Rebuilding society after a nuclear event is one of the most challenging tasks imaginable. The process involves not only restoring physical infrastructure but also rebuilding communities, governance systems, and social order. The psychological and social impact of such an event can be profound, requiring careful consideration of how to create a stable, resilient, and sustainable society in the aftermath. This chapter will explore the key steps, challenges, and strategies involved in rebuilding society after a nuclear event.

Establishing Immediate Governance and Order

Restoring Law and Order:

Interim Leadership: In the immediate aftermath of a nuclear event, establishing interim leadership is crucial for maintaining order and coordinating recovery efforts. This leadership can be based on existing governmental structures if they remain functional, or it may involve community leaders stepping into roles of responsibility. The focus should be on clear communication, decision-making, and maintaining public safety.

Emergency Laws and Regulations: Implement emergency laws and regulations that address the specific challenges of a post-nuclear environment. These might include curfews, rationing systems, and measures to prevent hoarding or looting. The goal is to maintain social order and ensure equitable distribution of resources.

Organizing Security Forces:

Local Security: Establish local security forces to protect people, property, and resources. These forces can be composed of surviving law enforcement, military personnel, or organized civilian volunteers. Their primary duties will include maintaining public order, preventing crime, and assisting with relief efforts.

Conflict Resolution: Develop systems for resolving conflicts that may arise over resources, territory, or other issues. This might involve creating a temporary judiciary or community-based dispute resolution mechanisms. Ensuring that conflicts are managed peacefully is essential for preventing further destabilization.

Restoring Communication and Information Networks

Re-establishing Communication Systems:

Radio Networks: In the absence of traditional communication infrastructure, setting up radio networks can provide a vital means of communication. These networks can be used for coordination among different groups, disseminating information, and broadcasting emergency updates.

Satellite Communication: If available, satellite communication systems can provide a more reliable and wide-reaching means of communication. These systems are less likely to be disrupted by the fallout and can be used to connect with other regions or international organizations.

Information Dissemination:

Public Information Campaigns: Start public information campaigns to keep the population informed about safety protocols, resource distribution, and recovery plans. Accurate and timely information is essential to prevent panic and ensure that people know what to do to stay safe.

Education and Training: Begin educating the population about the new realities of life in a post-nuclear world. This includes teaching basic survival skills, radiation safety, and the importance of community cooperation. Training programs should also be established to build skills that will be necessary for reconstruction and long-term survival.

Rebuilding Physical Infrastructure

Restoring Essential Services:

Water and Sanitation: The restoration of water and sanitation services is a top priority. This includes repairing or establishing new water supply systems, purifying water sources, and setting up sanitation facilities to prevent the spread of disease.

Power and Energy: Work to restore power supplies through whatever means are available, including repairing damaged infrastructure or deploying renewable energy systems like solar panels and wind turbines. Prioritize power for essential services such as hospitals, communication centers, and food storage facilities.

Transportation and Logistics:

Clearing Roads and Debris: Begin by clearing roads and removing debris to allow for the transportation of goods and people. This will facilitate the distribution of resources and aid in the movement of security forces and emergency services.

Re-establishing Supply Lines: Rebuild supply lines to ensure the flow of essential goods, such as food, water, medicine, and fuel. This may involve setting up temporary supply depots and using whatever vehicles or transportation methods are available.

Shelter and Housing:

Repairing or Building Shelters: Assess the availability and condition of existing shelters. Repair them as needed and construct new shelters if necessary. The immediate goal is to provide safe, radiation-protected housing for all survivors.

Long-Term Housing Solutions: Begin planning for longer-term housing solutions. This may involve rebuilding homes, constructing new residential areas, and using prefabricated or modular housing to quickly accommodate displaced populations.

Restoring the Economy and Food Supply

Economic Stabilization:

Barter and Trade: In the absence of a functioning monetary system, barter and trade will likely become the primary means of exchange. Establish local marketplaces and systems of trade that allow people to exchange goods and services. This will help to stabilize the economy and ensure that people have access to what they need.

Re-establishing Currency: As society stabilizes, work towards re-establishing a currency system. This could involve the use of existing currencies if they are still valid or the introduction of a new, locally backed currency. The goal is to facilitate broader economic activity and rebuilding efforts.

Reviving Agriculture:

Assessing Soil Safety: Test the soil for radiation contamination and assess its suitability for farming. Areas with lower contamination levels should be prioritized for agricultural use.

Starting Agriculture: Begin planting crops that are resilient and can be harvested quickly, such as root vegetables and grains. Focus on staple foods that provide essential nutrients and calories. If traditional farming is not possible, explore alternative methods such as hydroponics or indoor farming.

Securing the Food Supply:

Food Distribution: Set up food distribution centers to manage the equitable distribution of available food supplies. Prioritize the most vulnerable populations, such as children, the elderly, and the sick.

Stockpiling and Preservation: Begin stockpiling food supplies for future use. Focus on preserving foods through methods such as canning, drying, and smoking. This will help build a reserve of food that can be used in the event of future shortages.

Addressing Psychological and Social Recovery

Mental Health Support:

Psychological First Aid: Provide psychological first aid to help individuals cope with the trauma of the nuclear event. This can include counseling, support groups, and community activities that help people process their experiences and begin to heal.

Long-Term Mental Health Services: Establish long-term mental health services to address ongoing psychological issues such as post-traumatic stress disorder (PTSD), anxiety, and depression. These services are essential for helping people recover and rebuild their lives.

Rebuilding Social Structures:

Community Engagement: Encourage community engagement and participation in the rebuilding process. Involving people in decision-making and recovery efforts fosters a sense of ownership and helps rebuild social cohesion.

Restoring Cultural Practices: Support the restoration of cultural practices, traditions, and rituals that help people connect with their identity and community. This can play a vital role in social recovery and the rebuilding of a sense of normalcy.

Re-establishing Governance and Rule of Law

Transitioning from Emergency Governance:

Establishing a Permanent Government: As society stabilizes, transition from interim leadership to a more permanent form of governance. This may involve holding elections, re-establishing legislative bodies, and rebuilding government institutions.

Legal System and Justice: Restore the legal system to ensure the rule of law. This includes re-establishing courts, law enforcement, and correctional facilities. A functioning legal system is essential for maintaining order, resolving disputes, and protecting human rights.

International Relations and Diplomacy:

Reconnecting with the Global Community: Work to re-establish connections with the global community. This may involve reaching out to neighboring countries, international organizations, and humanitarian groups for assistance in the rebuilding process.

Negotiating Aid and Support: Engage in diplomatic efforts to secure aid, support, and resources from the international community. This can include financial aid, technical expertise, and access to global markets.

Planning for Future Resilience

Disaster Preparedness:

Developing Emergency Plans: Develop comprehensive emergency preparedness plans that address the potential for future nuclear events, as well as other disasters. These plans should include evacuation procedures, resource management strategies, and communication protocols.

Building Resilient Infrastructure: Focus on building infrastructure that is resilient to future disasters. This includes using durable materials, incorporating radiation shielding into new constructions, and planning for redundancy in critical systems like power, water, and communication.

Sustainability and Self-Sufficiency:

Resource Management: Implement sustainable resource management practices that ensure long-term self-sufficiency. This includes conserving water, managing agricultural resources, and developing renewable energy sources.

Education and Innovation: Invest in education and innovation to ensure that future generations are better prepared to face challenges. Encourage the development of new technologies, agricultural methods, and economic models that are suited to a post-nuclear world.

Cultural and Historical Preservation:

Documenting the Event: Preserve the history of the nuclear event and its aftermath by documenting survivor stories, rebuilding efforts, and lessons learned. This can be done through written records, oral histories, and digital archives.

Memorials and Remembrance: Establish memorials and spaces for remembrance to honor those who were lost and to serve as a reminder of the importance of peace and preparedness. These spaces can also foster a sense of community and shared history.

Rebuilding society after a nuclear event is a monumental challenge that requires coordinated efforts across multiple domains, including governance, infrastructure, economy, and social systems. The process involves not only addressing the immediate needs of survivors but also laying the groundwork for a stable, resilient, and sustainable future. By focusing on restoring essential services, re-establishing governance, supporting psychological and social recovery, and planning for future resilience, society can emerge stronger and more unified from the devastation of a nuclear event. The process of rebuilding will be long and arduous, but with determination, collaboration, and careful planning, it is possible to create a society that is not only functional but also better prepared to face future challenges.

The Role of Technology and Innovation in Rebuilding

Leveraging Technology for Reconstruction:

Advanced Building Techniques: Use innovative construction methods, such as 3D printing, prefabrication, and modular building, to quickly and efficiently rebuild infrastructure. These technologies can speed up the construction of homes, hospitals, and other critical facilities, ensuring that essential services are restored as soon as possible.

Digital Infrastructure: Rebuilding digital infrastructure, such as internet and telecommunications networks, is crucial for modern governance, education, and economic activity. Consider integrating smart technologies into new constructions to improve energy efficiency, security, and communication.

Agricultural Innovation:

Vertical Farming and Hydroponics: In areas where soil is contaminated, or space is limited, vertical farming and hydroponic systems can provide a sustainable solution for food production. These methods allow for the cultivation of crops indoors, using less water and avoiding soil contamination issues.

Genetically Modified Crops: Explore the use of genetically modified (GM) crops that are more resilient to radiation or harsh environmental conditions. These crops can help ensure food security in the challenging conditions of a post-nuclear world.

Energy Solutions:

Renewable Energy: Prioritize the development of renewable energy sources, such as solar, wind, and geothermal power, to reduce dependency on fossil fuels and create a more sustainable energy infrastructure. Renewable energy systems can be decentralized, making them more resilient to disruptions.

Energy Storage and Micro grids: Invest in energy storage solutions, like advanced batteries, and develop micro grids that can operate independently of the main grid. These systems ensure that essential services remain powered even if the main grid is compromised.

Cultural and Educational Revival

Restoring Education:

Rebuilding Schools: Re-establish educational institutions as soon as possible to provide stability and normalcy for children and youth. Schools can also serve as community hubs where knowledge is shared, and communal bonds are strengthened.

Adaptive Curricula: Develop curricula that address the new realities of a post-nuclear world, including survival skills, environmental stewardship, and community building. Education should also emphasize critical thinking, innovation, and adaptability.

Cultural Preservation and Innovation:

Reviving Arts and Culture: Encourage the revival of arts and cultural practices as a means of healing and rebuilding social cohesion. Art can be a powerful tool for processing trauma, expressing collective identity, and imagining a better future.

Innovation Hubs: Establish innovation hubs where people can collaborate on solutions to the unique challenges posed by the post-nuclear environment. These hubs can foster creativity, entrepreneurship, and the development of technologies that contribute to the rebuilding effort.

Global Collaboration and Long-Term Vision

International Aid and Collaboration:

Global Support Networks: Engage with international organizations, NGOs, and other countries to receive aid and share knowledge. Global support networks can provide critical resources, expertise, and funding that accelerate the rebuilding process.

Cross-Border Cooperation: Work with neighboring countries to address shared challenges, such as managing displaced populations, securing borders, and rebuilding trade routes. Cross-border cooperation can also help prevent conflicts over resources and ensure regional stability.

Vision for the Future:

Building a Resilient Society: Use the rebuilding process as an opportunity to create a society that is more resilient to future disasters. This includes building infrastructure that can withstand extreme events, fostering social cohesion, and developing robust governance systems that can adapt to changing conditions.

Promoting Peace and Disarmament: The experience of rebuilding after a nuclear event should reinforce the importance of global efforts toward nuclear disarmament and non-proliferation. Advocating for peace and the responsible use of nuclear technology can help prevent future catastrophes.

Sustaining Hope and Motivation:

Shared Vision and Goals: Develop a shared vision for the future that inspires and motivates the population. This vision should be communicated clearly and consistently, emphasizing the progress that has been made and the collective goals for rebuilding.

Celebrating Milestones: As society rebuilds, celebrate milestones and achievements, no matter how small. Recognizing progress can boost morale, foster a sense of accomplishment, and encourage continued efforts toward recovery.

Rebuilding society after a nuclear event is an immense undertaking that demands coordination, resilience, and a forward-looking vision. By addressing the immediate needs for governance, infrastructure, and security, while also laying the foundation for long-term recovery and sustainability, society can rise from the ashes of disaster. The process will be marked by challenges, setbacks, and difficult decisions, but through innovation, collaboration, and a commitment to shared goals, it is possible to create a society that is not only restored but transformed—stronger, more resilient, and more aware of the importance of peace and preparedness.

Lessons from History: Past Nuclear Events and Their Impact

Understanding the lessons from past nuclear events is essential for preparing for, responding to, and recovering from potential future nuclear disasters. History provides valuable insights into the consequences of nuclear events, the effectiveness of responses, and the long-term impact on societies and the environment. This chapter will examine key nuclear events in history, their immediate and long-term effects, and the lessons they offer for building a safer and more resilient future.

Hiroshima and Nagasaki: The First Use of Nuclear Weapons

Background and Context:

The Atomic Bombings: On August 6 and 9, 1945, the United States dropped atomic bombs on the Japanese cities of Hiroshima and Nagasaki. These bombings marked the first and only use of nuclear weapons in war, leading to the immediate deaths of over 200,000 people and causing widespread destruction.

End of World War II: The bombings played a significant role in Japan's surrender, effectively ending World War II. However, they also ushered in the nuclear age, highlighting the devastating power of nuclear weapons and setting the stage for the Cold War.

Immediate Impact:

Destruction and Casualties: The bombings caused unprecedented destruction, leveling entire cities and killing tens of thousands instantly. Many more died in the following days and weeks from radiation sickness, burns, and other injuries.

Radiation Effects: Survivors, known as hibakusha, suffered from acute radiation sickness, characterized by nausea, vomiting, hair loss, and a weakened immune system. Long-term effects included an increased risk of cancer, birth defects, and other health issues.

Long-Term Impact:

Reconstruction and Recovery: Both Hiroshima and Nagasaki underwent extensive reconstruction in the post-war years. Despite the challenges, the cities were rebuilt, and today they serve as symbols of peace and the human capacity for resilience.

Nuclear Disarmament Advocacy: The bombings spurred global movements advocating for nuclear disarmament. Survivors and activists have worked tirelessly to raise awareness of the horrors of nuclear war and to promote the elimination of nuclear weapons.

Lessons Learned:

Humanitarian Consequences: The bombings of Hiroshima and Nagasaki underscore the catastrophic humanitarian consequences of nuclear weapons. The immense loss of life, long-term health effects, and environmental damage serve as a stark reminder of the need to prevent the use of nuclear weapons.

Importance of Peacebuilding: The events highlight the importance of diplomatic efforts and peacebuilding to resolve conflicts and avoid the use of nuclear weapons. International cooperation and dialogue are essential to reducing the risk of nuclear war.

The Chernobyl Disaster: A Nuclear Accident with Global Repercussions

Background and Context:

The Accident: On April 26, 1986, a reactor at the Chernobyl Nuclear Power Plant in the Soviet Union (now Ukraine) exploded during a safety test, releasing massive amounts of radioactive material into the atmosphere. It is considered the worst nuclear disaster in history.

Evacuation and Containment: The nearby city of Pripyat was evacuated, and a 30-kilometer exclusion zone was established around the plant. Soviet authorities launched a massive containment effort, including the construction of a sarcophagus over the damaged reactor.

Immediate Impact:

Radiation Release: The explosion released radioactive isotopes, including iodine-131, cesium-137, and strontium-90, which spread across Europe. The initial release was estimated to be hundreds of times greater than the atomic bombs dropped on Hiroshima and Nagasaki.

Health Consequences: Thousands of people were exposed to high levels of radiation, leading to acute radiation sickness and a significant increase in thyroid cancer, particularly among children. Many first responders, known as liquidators, suffered severe health consequences due to their exposure during the containment efforts.

Long-Term Impact:

Environmental Damage: The Chernobyl disaster had long-lasting environmental effects, contaminating vast areas of land and water. The exclusion zone remains largely uninhabited, though it has become a unique ecological reserve.

Socio-political Fallout: The disaster eroded public trust in the Soviet government and contributed to the eventual collapse of the Soviet Union. It also led to significant changes in nuclear policy and safety regulations worldwide.

Lessons Learned:

Nuclear Safety and Transparency: The Chernobyl disaster highlighted the importance of strict nuclear safety protocols and the need for transparency in nuclear operations. International cooperation and independent oversight are crucial to preventing and responding to nuclear accidents.

Public Health Preparedness: The disaster underscored the need for robust public health preparedness in the event of nuclear accidents. This includes early warning systems, rapid evacuation plans, and the availability of medical care and iodine tablets to mitigate radiation exposure.

The Fukushima Daiichi Disaster: Modern Nuclear Challenges

Background and Context:

The Earthquake and Tsunami: On March 11, 2011, a massive earthquake and tsunami struck Japan, leading to the failure of cooling systems at the Fukushima Daiichi Nuclear Power Plant. This caused three reactors to melt down, releasing radioactive material into the environment.

Evacuation and Response: Approximately 160,000 people were evacuated from the area, and a 20-kilometer exclusion zone was established. The Japanese government and TEPCO (Tokyo Electric Power Company) faced criticism for their handling of the disaster, particularly in the early stages.

Immediate Impact:

Radiation Release: The disaster released significant amounts of radioactive material, including iodine-131 and cesium-137, into the air and sea. Contaminated water and soil posed ongoing challenges for clean-up efforts.

Health and Safety: While there were no immediate deaths from radiation exposure, the disaster had serious health impacts, including increased stress, anxiety, and the potential for long-term cancer risks among those exposed to lower levels of radiation.

Long-Term Impact:

Environmental and Economic Costs: The Fukushima disaster caused widespread environmental contamination, particularly of the marine environment. The cleanup and decommissioning of the plant are expected to take decades and cost billions of dollars.

Impact on Global Nuclear Policy: The disaster led to a global reassessment of nuclear energy policies. Several countries, including Germany, decided to phase out nuclear power, while others implemented stricter safety regulations and emergency preparedness measures.

Lessons Learned:

Natural Disaster Preparedness: Fukushima highlighted the vulnerability of nuclear power plants to natural disasters. It emphasized the need for robust disaster preparedness plans that account for extreme events, including earthquakes, tsunamis, and other environmental hazards.

Crisis Communication and Trust: The importance of clear, timely communication during a nuclear crisis was underscored by the Fukushima disaster. Building and maintaining public trust through transparency and effective communication is essential for managing the aftermath of such events.

The Three Mile Island Incident: A Near-Disaster in the United States

Background and Context:

The Incident: On March 28, 1979, a partial meltdown occurred in one of the reactors at the Three Mile Island Nuclear Generating Station in Pennsylvania, USA. Although the incident was contained without significant release of radiation, it caused widespread public concern.

Response and Containment: The incident led to the evacuation of pregnant women and young children from the area as a precaution. The cleanup and decommissioning of the damaged reactor took several years and cost over a billion dollars.

Immediate Impact:

Public Panic: The incident triggered widespread public panic and fear of a major nuclear disaster. Miscommunication and conflicting reports from authorities exacerbated the situation, leading to a loss of public confidence in nuclear energy.

Regulatory Changes: In response to the incident, the U.S. Nuclear Regulatory Commission (NRC) implemented significant changes to nuclear safety regulations, including improvements in reactor design, operator training, and emergency preparedness.

Long-Term Impact:

Shift in Public Opinion: The Three Mile Island incident significantly shifted public opinion against nuclear power in the United States. It led to the cancellation of numerous nuclear projects and a slowdown in the development of new nuclear plants.

Nuclear Industry Reforms: The incident prompted the nuclear industry to adopt a culture of safety and continuous improvement. This included the establishment of the Institute of Nuclear Power Operations (INPO) to promote safety and excellence in nuclear operations.

Lessons Learned:

Human Error and Training: The incident highlighted the role of human error in nuclear safety and the importance of rigorous training for nuclear plant operators. Ensuring that personnel are well-trained and that systems are designed to minimize the risk of human error is critical for preventing nuclear accidents.

Emergency Preparedness and Public Communication: The Three Mile Island incident demonstrated the need for robust emergency preparedness plans and effective public communication strategies. Authorities must be prepared to respond quickly and transparently to prevent panic and ensure public safety.

The Cold War and Nuclear Proliferation: A Global Threat

Background and Context:

The Nuclear Arms Race: During the Cold War, the United States and the Soviet Union engaged in a nuclear arms race, amassing vast arsenals of nuclear weapons. The threat of mutually assured destruction (MAD) kept both superpowers in check but also created a constant risk of nuclear war.

Nuclear Proliferation: The spread of nuclear weapons to other countries, including the United Kingdom, France, China, and later India, Pakistan, and North Korea, increased the complexity of global security dynamics. Efforts to prevent proliferation led to the establishment of international treaties and organizations.

Impact on Global Security:

The Cuban Missile Crisis: The Cuban Missile Crisis in 1962 brought the world to the brink of nuclear war. The crisis highlighted the dangers of nuclear brinkmanship and the importance of diplomacy and communication in preventing nuclear conflict. The successful resolution of the crisis through diplomatic negotiations underscored the necessity of dialogue and de-escalation in managing nuclear tensions.

Nuclear Accidents and Near Misses: Throughout the Cold War, several incidents, including false alarms and accidental launches, nearly led to unintended nuclear exchanges. These near misses demonstrated the fragility of nuclear deterrence and the potential for catastrophic mistakes.

Long-Term Impact:

Arms Control Treaties: The Cold War saw the development of several key arms control treaties aimed at reducing the risk of nuclear war and curbing the proliferation of nuclear weapons. The Treaty on the Non-Proliferation of Nuclear Weapons (NPT), Strategic Arms Limitation Talks (SALT), and later the Strategic Arms Reduction Treaty (START) were instrumental in limiting the nuclear arms race and promoting global security.

Nuclear Non-Proliferation and Disarmament: The legacy of the Cold War continues to shape global efforts toward nuclear non-proliferation and disarmament. The fear of nuclear war and the recognition of the devastating consequences of nuclear weapons have fueled international movements and treaties aimed at reducing and eventually eliminating nuclear arsenals.

Lessons Learned:

The Importance of Diplomacy: The Cold War highlighted the critical role of diplomacy in preventing nuclear conflict. Maintaining open lines of communication, building trust between nuclear powers, and engaging in arms control negotiations are essential for global security.

Managing Nuclear Risks: The history of the Cold War underscores the need for robust systems to manage nuclear risks, including strict command and control structures, secure communication channels, and fail-safes to prevent accidental launches. Continuous efforts to improve these systems are necessary to reduce the likelihood of a nuclear catastrophe.

Lessons from Lesser-Known Nuclear Incidents

The Kyshtym Disaster (1957):

Background: The Kyshtym disaster occurred in 1957 at the Mayak Production Association in the Soviet Union. A cooling system failure led to the explosion of a storage tank containing radioactive waste, resulting in significant radioactive contamination of the surrounding area.

Impact: The disaster exposed thousands of people to high levels of radiation, and entire villages had to be evacuated. However, the Soviet government kept the incident secret for many years, hindering proper response and recovery efforts.

Lessons Learned: The Kyshtym disaster highlights the dangers of secrecy and lack of transparency in handling nuclear incidents. Effective disaster response requires timely public communication, international cooperation, and accountability.

The Goiânia Accident (1987):

Background: The Goiânia accident in Brazil involved the improper disposal of a radioactive medical device containing cesium-137. The device was found by scavengers, who unknowingly spread the radioactive material, resulting in severe contamination and multiple deaths.

Impact: The accident led to widespread panic, the evacuation of affected areas, and the contamination of hundreds of people. The clean-up took several months, and the incident had long-term health and psychological effects on the affected population.

Lessons Learned: The Goiânia accident underscores the importance of proper disposal and handling of radioactive materials. Public education on the dangers of radiation and strict regulations for managing radioactive waste are essential to prevent similar incidents.

The Windscale Fire (1957):

Background: The Windscale fire occurred at a nuclear reactor in the United Kingdom. A fire broke out during an experiment, releasing significant amounts of radioactive material into the atmosphere.

Impact: The fire led to the contamination of surrounding areas and required extensive clean-up efforts. The incident raised concerns about the safety of nuclear reactors and led to changes in reactor design and safety protocols.

Lessons Learned: The Windscale fire highlights the importance of rigorous safety procedures and the need for continuous monitoring and maintenance of nuclear facilities. The incident also demonstrated the need for effective emergency response plans and the importance of learning from accidents to improve safety standards.

The Role of International Organizations in Nuclear Safety

The International Atomic Energy Agency (IAEA):

Mandate and Role: The IAEA plays a central role in promoting the safe, secure, and peaceful use of nuclear technology. It establishes international safety standards, conducts inspections, and provides assistance to countries in managing nuclear facilities and responding to emergencies.

Impact: The IAEA's work has been instrumental in improving nuclear safety worldwide. Its efforts in response to incidents like Chernobyl and Fukushima have helped to coordinate international support, improve safety practices, and facilitate the sharing of knowledge and best practices.

Lessons Learned: The IAEA's work underscores the importance of international cooperation and oversight in nuclear safety. Global challenges require global solutions, and the IAEA's role in fostering collaboration and setting standards is critical to preventing nuclear accidents and promoting the peaceful use of nuclear technology.

The Comprehensive Nuclear-Test-Ban Treaty Organization (CTBTO):

Mandate and Role: The CTBTO oversees the implementation of the Comprehensive Nuclear-Test-Ban Treaty (CTBT), which prohibits all nuclear explosions. The organization operates a global monitoring system to detect nuclear tests and promote the treaty's entry into force.

Impact: The CTBTO's monitoring system has been successful in detecting nuclear tests and contributing to global security by deterring nuclear testing. The treaty, once fully ratified, will be a significant step toward nuclear disarmament.

Lessons Learned: The CTBTO's work highlights the importance of verification and monitoring in enforcing international agreements. A robust and transparent monitoring system is essential for building trust and ensuring compliance with global non-proliferation efforts.

The Humanitarian Impact of Nuclear Events

Health Effects and Long-Term Consequences:

Radiation Sickness and Cancer: Exposure to high levels of radiation can cause acute radiation sickness, which includes symptoms such as nausea, vomiting, and fatigue. Long-term exposure increases the risk of cancer, particularly thyroid cancer, leukemia, and other malignancies.

Genetic and Reproductive Effects: Radiation exposure can lead to genetic mutations, which may be passed on to future generations. Additionally, exposure can affect fertility and lead to complications in pregnancy and childbirth.

Mental Health and Trauma: Survivors of nuclear events often experience severe psychological trauma, including post-traumatic stress disorder (PTSD), anxiety, and depression. The loss of loved ones, displacement, and the stigma associated with radiation exposure can exacerbate these issues.

Environmental and Societal Impact:

Ecosystem Damage: Nuclear events can cause long-term damage to ecosystems, contaminating soil, water, and air. This contamination can lead to the death of plants and animals, disrupt food chains, and make large areas uninhabitable.

Economic Disruption: The economic impact of a nuclear event can be devastating, including the loss of infrastructure, disruption of trade, and the long-term costs of cleanup and decommissioning. Rebuilding affected areas can take decades and require significant financial resources.

Displacement and Social Disruption: Nuclear events often lead to the displacement of large populations, disrupting communities and social structures. Rebuilding social cohesion and providing support for displaced individuals and families is a critical part of the recovery process.

The history of nuclear events offers crucial lessons for the future. From the devastation of Hiroshima and Nagasaki to the long-lasting impact of Chernobyl and Fukushima, these events highlight the need for stringent safety measures, transparent communication, and international cooperation. They also underscore the importance of nuclear disarmament and the prevention of nuclear conflict.

As we look to the future, it is essential to apply these lessons to build a safer, more resilient world. This includes continuing to improve nuclear safety standards, strengthening global non-proliferation efforts, and promoting the peaceful use of nuclear technology. By learning from the past and working together, we can reduce the risks associated with nuclear energy and weapons, and ensure that the horrors of past nuclear events are never repeated.

How to Stay Informed About Nuclear Threats

Staying informed about nuclear threats is crucial for ensuring your safety and preparedness in an increasingly complex world. With advancements in technology and shifts in global geopolitics, the landscape of nuclear risks continues to evolve. By understanding how to stay informed and what sources to rely on, you can make timely and informed decisions about how to protect yourself and your community. This chapter will guide you through the various ways to stay informed about nuclear threats, including monitoring reliable sources, using technology, and understanding the signals that indicate a heightened risk.

Understanding the Types of Nuclear Threats

Nuclear Weapons Threats:

State-Sponsored Nuclear Weapons: These threats come from nations with nuclear arsenals. Understanding the geopolitical tensions and relations between nuclear-armed states can provide insight into potential risks.

Nuclear Terrorism: This involves non-state actors, such as terrorist organizations, acquiring and potentially using nuclear materials. Staying informed about the activities and capabilities of these groups is essential for assessing this type of threat.

Nuclear Accidents and Incidents:

Nuclear Power Plant Accidents: Accidents at nuclear power plants can release dangerous levels of radiation. Understanding the safety measures in place at nearby plants and staying updated on their operational status is important.

Nuclear Material Theft or Smuggling: The illegal trade and theft of nuclear materials pose a significant threat. Being aware of reports and alerts regarding the movement of these materials can help you assess the risk of nuclear terrorism or accidents.

Reliable Sources of Information

Government Agencies:

National Emergency Management Agencies: Agencies like FEMA in the United States or the UK's Civil Contingencies Secretariat provide timely information and alerts about nuclear threats. They often issue public guidance and emergency instructions in the event of a crisis.

Defense and Security Departments: Departments such as the U.S. Department of Defense or the Russian Ministry of Defence may release statements or reports on national security threats, including those related to nuclear weapons.

International Organizations:

International Atomic Energy Agency (IAEA): The IAEA is a leading authority on nuclear safety and security. It monitors nuclear activities worldwide and provides updates on nuclear incidents, safety standards, and the status of nuclear facilities.

World Health Organization (WHO): In the event of a nuclear accident or radiation release, the WHO provides health-related information, including guidelines on radiation exposure and medical treatments.

News Outlets and Media:

Reputable News Agencies: Trusted news organizations like BBC, Reuters, The Associated Press, and Al Jazeera provide timely and accurate reporting on global nuclear threats. These outlets have established credibility and often have correspondents in key locations.

Specialized Security Publications: Publications such as *Jane's Defence Weekly*, *The Bulletin of the Atomic Scientists*, and *Arms Control Today* offer in-depth analysis and updates on nuclear security issues.

Non-Governmental Organizations (NGOs):

Nuclear Threat Initiative (NTI): NTI provides resources and updates on nuclear and biological threats. Their work includes tracking nuclear security incidents and providing policy recommendations.

Global Zero: An organization focused on the elimination of nuclear weapons, Global Zero offers insights into disarmament efforts and the current state of global nuclear arsenals.

Technology and Tools for Staying Informed

Mobile Apps and Alerts:

Emergency Alert Systems: Many countries have mobile apps or SMS services that provide emergency alerts for nuclear threats. For example, in the U.S., the FEMA app and Wireless Emergency Alerts (WEA) system can send notifications about imminent threats.

Radiation Monitoring Apps: Apps like RadResponder or the Safecast app allow you to monitor local radiation levels and receive alerts about changes in environmental radiation.

Social Media and Online Platforms:

Twitter and Social Media Feeds: Following government agencies, international organizations, and reputable news outlets on social media can provide real-time updates on nuclear threats. Twitter, in particular, is used by many organizations for rapid communication.

Online Forums and Communities: Platforms like Reddit have communities dedicated to emergency preparedness, where users share information, updates, and resources related to nuclear threats. However, always verify information from such sources with trusted authorities.

Geolocation Tools:

Nuclear Facilities Maps: Tools like the Nuclear Facilities Map from the IAEA or the NRDC's Nuclear Data Mapping Project provide locations of nuclear power plants, weapons sites, and storage facilities worldwide. Understanding your proximity to these sites can help you assess risk.

Global Incident Maps: Websites like the Global Incident Map track and display various global events, including nuclear incidents, providing a visual overview of potential threats around the world.

Monitoring Global Events and Indicators

Geopolitical Tensions:

Understanding Flashpoints: Certain regions, such as the Korean Peninsula, South Asia, or the Middle East, are considered nuclear flashpoints due to ongoing conflicts and the presence of nuclear weapons. Monitoring developments in these areas is crucial for staying informed about potential nuclear risks.

Diplomatic Signals: Pay attention to diplomatic developments, such as treaty negotiations, summits, or breakdowns in dialogue between nuclear-armed states. These events can indicate shifts in nuclear policy or the likelihood of conflict.

Nuclear Testing and Military Exercises:

Detection of Nuclear Tests: Organizations like the Comprehensive Nuclear-Test-Ban Treaty Organization (CTBTO) monitor for nuclear tests globally. Reports of detected tests can signal increased tensions or developments in nuclear capabilities.

Military Drills: Large-scale military exercises by nuclear-armed states can be a sign of escalating tensions. Tracking these exercises, especially those involving nuclear-capable forces, can provide insights into the security environment.

Environmental Monitoring:

Radiation Detection Networks: Networks like the RadNet system in the U.S. or Europe's Ring of Five nuclear radiation monitoring network continuously monitor radiation levels. Abnormal spikes in radiation can indicate an accident, test, or other nuclear incident.

Global Atmospheric Models: After a nuclear event, atmospheric dispersion models can predict the spread of radioactive materials. Websites and apps that provide access to these models can help you understand the potential impact on your location.

Staying Prepared and Responsive

Personal and Community Preparedness:

Emergency Plans: Develop and regularly update a personal or family emergency plan that includes steps to take in the event of a nuclear threat. This should cover evacuation routes, communication plans, and sheltering options.

Community Involvement: Engage with local emergency preparedness groups or initiatives. Being part of a community effort can enhance your access to information and resources during a nuclear emergency.

Education and Training:

Radiation Safety Training: Consider taking courses or participating in training programs that cover radiation safety, first aid for radiation exposure, and general emergency preparedness. Knowing how to respond can significantly reduce your risk.

Regular Drills: Conduct regular drills at home or in your community to ensure that everyone knows what to do in the event of a nuclear threat. Practice makes it easier to act quickly and effectively when a real threat arises.

Psychological Preparedness:

Staying Calm and Rational: In the event of a nuclear threat, it's crucial to stay calm and think clearly. Being well-informed helps reduce anxiety and allows you to make better decisions under pressure.

Mental Health Resources: Access mental health resources to help cope with the stress and anxiety that may arise from living under the threat of nuclear incidents. Counseling and support groups can be invaluable for maintaining mental well-being.

Navigating Information Overload and Misinformation

Critical Evaluation of Sources:

Assessing Credibility: Always verify information against multiple sources before acting on it. Trusted government and international organizations are usually the most reliable sources of information.

Misinformation and Hoaxes: Be aware that misinformation and hoaxes can spread rapidly during a crisis. Approach sensational claims with skepticism and rely on information from official channels.

Using Trusted Networks:

Building Information Networks: Establish a network of trusted contacts who can provide reliable information during a nuclear threat. This might include local emergency services, community leaders, and knowledgeable friends or family members.

Verifying through Official Channels: Always cross-check information with official channels, such as government websites, official press releases, or direct communications from recognized authorities.

Staying informed about nuclear threats is an essential part of preparedness in today's world. By understanding the types of threats, monitoring reliable sources, and using available technology, you can stay ahead of potential risks and take proactive steps to protect yourself and your loved ones.

Remaining vigilant, staying calm, and regularly updating your knowledge and preparedness plans will help you navigate the complex and evolving landscape of nuclear threats.

Educating Children about Nuclear Safety

Educating children about nuclear safety is an essential part of preparing them for the unlikely but serious possibility of a nuclear event. It's important to approach this topic with care, balancing the need for information with the need to avoid causing unnecessary fear. By teaching children about nuclear safety in a calm, clear, and age-appropriate manner, you can help them understand the importance of preparedness and ensure they know what to do in an emergency. This chapter will guide you through effective strategies for educating children about nuclear safety, including age-appropriate explanations, practical training, and psychological support.

Introducing Nuclear Safety: Age-Appropriate Explanations

Early Childhood (Ages 3-7):

Simple Concepts: For younger children, focus on the basics of safety without delving into complex or frightening details. Explain that there are certain things in the world, like storms or bad weather, that we need to be prepared for, and sometimes this includes something called "radiation" or "nuclear."

Safety Rules: Teach simple safety rules, such as staying indoors when told, listening to adults during emergencies, and knowing basic steps like covering their nose and mouth if instructed. Use stories or cartoons that illustrate the importance of following safety rules in a non-scary way.

Emergency Practice: Introduce the concept of practice drills in a fun and engaging manner, like playing a game. For example, you can pretend to be at school or home and practice what to do if an emergency happens, emphasizing that it's just practice and there's nothing to worry about.

Middle Childhood (Ages 8-12):

Understanding Radiation: Children in this age group can begin to understand the concept of radiation as something that can make people sick if they are exposed to too much of it. Use analogies, like comparing it to the sun's rays, which are mostly safe but can be harmful if you're out too long without protection.

Nuclear Events: Explain that sometimes, there are accidents or emergencies where this harmful radiation could be in the air, but there are ways to stay safe, like going inside quickly, staying away from windows, and following instructions from adults or emergency workers.

Building Confidence: Emphasize that by knowing what to do, they are helping to keep themselves and their family safe. Practice simple emergency steps, like how to quickly get indoors and where to go in the house for safety, in a way that makes them feel empowered rather than scared.

Adolescence (Ages 13-18):

Detailed Information: Teens can handle more detailed explanations about nuclear safety. Discuss the science behind radiation, the potential sources of nuclear threats (like power plant accidents or nuclear weapons), and why certain safety measures are necessary.

Critical Thinking: Encourage critical thinking by discussing the importance of reliable information and how to recognize credible sources. Talk about the role of government agencies and emergency services in managing nuclear safety and how they can stay informed through news and official alerts.

Practical Preparedness: Involve teens in more practical aspects of preparedness, such as assembling emergency kits, understanding evacuation routes, and learning basic first aid. Empower them to take an active role in family safety plans, which can help reduce anxiety by giving them a sense of control.

Practical Training and Drills

Home Safety Drills:

Designing Drills: Create home safety drills that simulate what to do in the event of a nuclear emergency. This could involve practicing how to quickly move to a designated safe area in the home, such as a basement or interior room, where they would be shielded from radiation.

Making it Routine: Regularly practice these drills, similar to fire drills, so that children know exactly what to do without panic. Reinforce that these drills are just practice to help everyone stay safe and that there's no immediate danger.

School Involvement:

School Safety Programs: Ensure that your child's school has a nuclear safety plan and participates in regular drills. Work with school administrators to understand their protocols and ensure they align with what you are teaching at home.

Coordinated Efforts: Encourage schools to involve parents in drills and provide information on how they are teaching students about nuclear safety. This helps ensure consistent messaging and preparedness between home and school.

Public Awareness Campaigns:

Community Drills: Participate in or organize community-wide drills that involve both children and adults. These events can be a way to educate children in a larger context and make the concept of preparedness a normal part of community life.

Educational Materials: Use materials provided by local emergency management agencies, which are often designed to be child-friendly, to supplement your teaching. These can include booklets, videos, and interactive games that reinforce key safety concepts.

Psychological Support and Reassurance

Open Communication:

Encouraging Questions: Create an environment where children feel comfortable asking questions about nuclear safety. Answer their questions honestly but in an age-appropriate way. If you don't know the answer, look it up together, which also teaches them how to find reliable information.

Addressing Fears: Acknowledge any fears they may have and provide reassurance. Let them know that while it's normal to feel scared about things we don't fully understand, being prepared helps everyone stay safe. Emphasize the low likelihood of a nuclear event but the importance of being ready just in case.

Reducing Anxiety:

Focusing on Safety: Shift the focus from the dangers of nuclear threats to the safety measures that can be taken. Reinforce that many people, such as scientists, government officials, and emergency responders, are working hard to keep everyone safe.

Normalizing Preparedness: Integrate nuclear safety education into a broader context of general emergency preparedness, so it doesn't seem overly alarming. Discussing it alongside other safety topics like fire drills, earthquake preparedness, or severe weather safety can help make it seem like just another aspect of being safe.

Positive Reinforcement:

Celebrating Preparedness: After practicing drills or discussing nuclear safety, praise your child for their participation and understanding. Reinforcing their efforts with positive feedback helps build their confidence and resilience.

Engaging Activities: Use creative activities like drawing, role-playing, or building an emergency kit together to make learning about nuclear safety more engaging and less intimidating.

Resources for Further Education

Books and Educational Materials:

Child-Friendly Books: There are children's books designed to explain nuclear safety in a way that is easy to understand and reassuring. Look for titles recommended by educators or public safety organizations.

Interactive Learning: Explore interactive online resources or apps that teach nuclear safety through games or simulations. These tools can make learning about safety fun and memorable.

Community and School Programs:

Workshops and Classes: Some communities offer workshops or classes on emergency preparedness that are suitable for children. These programs often cover a range of safety topics, including nuclear safety.

Scout and Youth Programs: Organizations like the Boy Scouts or Girl Scouts may have badges or activities related to emergency preparedness. These programs can be a valuable way for children to learn about nuclear safety in a structured and supportive environment.

Adapting Education to Special Needs

Tailoring to Individual Needs:

Special Needs Considerations: If your child has special needs, adapt the nuclear safety education to their level of understanding and ability. This might involve breaking down information into smaller, more manageable parts or using more visual aids.

Professional Support: Consult with educators or therapists who work with your child to develop strategies that will be most effective. They can offer guidance on how to present information in a way that is both accessible and reassuring.

Reinforcing Through Routine:

Consistent Practice: Children with special needs may benefit from more frequent practice and repetition of safety drills. Consistent routines can help them feel more secure and confident in knowing what to do in an emergency.

Positive Reinforcement: Use positive reinforcement techniques to encourage participation and understanding. Celebrate successes, no matter how small, to build confidence and reduce anxiety.

Educating children about nuclear safety is a delicate but essential task that can be approached in a way that is both informative and reassuring. By tailoring the information to their age and development level, engaging them in practical training, and providing ongoing psychological support, you can help them understand the importance of nuclear safety without instilling unnecessary fear.

Remember, the goal is to empower children with the knowledge and skills they need to stay safe in a nuclear emergency, while also fostering a sense of security and confidence in their ability to handle the situation.

Protecting Pets in a Fallout Scenario

In a fallout scenario, ensuring the safety and well-being of your pets is a vital aspect of your emergency preparedness plan. Pets are vulnerable to radiation exposure just like humans, and they rely on you to protect them during such an emergency. This chapter will provide guidance on how to protect your pets in a fallout scenario, including preparation steps, sheltering strategies, and post-fallout care.

Preparing for a Fallout Scenario with Pets

Emergency Kit for Pets:

Food and Water: Stockpile at least two weeks' worth of food and water specifically for your pets. Make sure the food is stored in airtight, waterproof containers to prevent contamination. Include collapsible bowls for feeding.

Medications and Veterinary Records: If your pet requires any medications, ensure you have a sufficient supply stored in your emergency kit. Also, include a copy of your pet's vaccination records, any important medical history, and contact information for your veterinarian.

Pet First Aid Kit: Assemble a first aid kit that includes bandages, antiseptics, tweezers, scissors, and any other supplies recommended by your veterinarian. A pet-specific first aid manual can also be helpful.

Comfort Items: Include a few items that can help reduce your pet's stress, such as a favorite toy, blanket, or bed. Familiar items can provide comfort during a stressful situation.

Shelter Preparation:

Identifying Safe Zones: Identify a safe area in your home where you and your pets can shelter during fallout. This should be an interior room, basement, or any location away from windows and exterior walls to minimize exposure to radiation.

Pet-Friendly Shelter: Make sure your shelter area is pet-friendly. This means providing space for your pet to move around, designated areas for feeding, and a place for them to relieve themselves. For small pets, consider using pet carriers to keep them secure during the initial stages of the fallout.

Litter and Waste Disposal: For cats, ensure you have a portable litter box and enough litter to last at least two weeks. For dogs, stock up on disposable waste bags. Proper waste disposal is crucial to maintain hygiene in your shelter.

Evacuation Planning:

Pet-Friendly Evacuation Centers: Identify pet-friendly evacuation centers or shelters in advance. Not all emergency shelters accept pets, so knowing where you can go with your pets is essential.

Transporting Your Pet: Ensure you have carriers for small pets and leashes or harnesses for larger ones. Practice loading and unloading your pets into vehicles to reduce stress during an actual evacuation.

Emergency Contacts: Have a list of emergency contacts, including nearby friends or family members who could help care for your pets if you are unable to.

Sheltering with Pets during Fallout

Bringing Pets Inside:

Immediate Action: As soon as you are aware of a fallout event, bring your pets indoors immediately. Do not allow them to remain outside where they could be exposed to radioactive particles.

Decontamination: Before bringing pets into your shelter area, brush off their fur to remove any possible radioactive dust. If possible, wash them with soap and water to reduce contamination. Wear gloves and protective clothing while doing this to protect yourself from exposure.

Maintaining a Safe Shelter Environment:

Air Quality: Ensure proper ventilation in your shelter area, but avoid opening windows or doors that could let in contaminated air. Consider using an air purifier with a HEPA filter to help reduce airborne contaminants.

Minimizing Stress: Pets can sense stress and anxiety, which can affect their behavior. Keep the environment as calm as possible by speaking softly and maintaining a routine that includes feeding, play, and rest times.

Limiting Exposure: Limit your pet's exposure to the outside environment during and immediately after the fallout. Keep them in the designated safe area and avoid letting them wander around the house where they could come into contact with radioactive materials.

Handling Pet Waste:

Hygiene: Maintain hygiene in your shelter by regularly cleaning up after your pets. Dispose of waste in sealed plastic bags and store them away from the shelter area until it is safe to dispose of them outside.

Litter Boxes and Pee Pads: For cats, regularly scoop the litter box to keep it clean. For dogs or other pets, use pee pads that can be easily replaced to maintain cleanliness in the shelter.

Post-Fallout Care for Pets

Assessing the Environment:

Radiation Monitoring: Before allowing your pets back outside, use a radiation detector to check for residual radiation levels in your area. Only let your pets out once it is safe to do so.

Decontaminating Your Home: After the fallout has settled, clean your home thoroughly to remove any radioactive particles. Pay special attention to areas where your pets spend a lot of time. Vacuum carpets, wipe down surfaces, and wash bedding and toys.

Health Monitoring:

Watch for Symptoms: Monitor your pets for any signs of radiation sickness, such as vomiting, diarrhoea, loss of appetite, fatigue, or unusual behavior. If you notice any of these symptoms, seek veterinary care immediately.

Regular Vet Check-ups: After a fallout event, schedule a veterinary check-up as soon as possible to ensure your pet has not been adversely affected. Even if your pet appears healthy, a professional examination is important.

Reintroducing Pets to the Outdoors:

Gradual Reintroduction: When it is safe to do so, reintroduce your pets to the outdoors gradually. Start with short, supervised outings to ensure they do not come into contact with any lingering radioactive materials.

Protective Measures: If radiation levels are still slightly elevated, consider using protective clothing for your pets, such as dog boots or a lightweight coat, to minimize contact with contaminated surfaces.

Special Considerations for Different Types of Pets

Small Animals (Rabbits, Guinea Pigs, etc.):

Sheltering: Small animals should be kept in their cages or carriers within the shelter area to keep them safe and contained. Cover their cages with a cloth to reduce stress and provide a sense of security.

Ventilation: Ensure adequate ventilation in their cages but avoid direct exposure to drafts or contaminated air.

Birds:

Air Quality: Birds are particularly sensitive to air quality, so ensure that your shelter is well-ventilated but free of contaminated air. Use an air purifier if possible.

Stress Reduction: Birds can be easily stressed by changes in their environment. Cover their cage with a breathable cloth to provide comfort and reduce anxiety.

Reptiles and Amphibians:

Temperature Control: Reptiles and amphibians require specific temperature and humidity levels to remain healthy. Make sure you can maintain these conditions in your shelter area.

Handling and Containment: Keep reptiles and amphibians in their terrariums or enclosures during the fallout event. Avoid handling them excessively to reduce stress.

Long-Term Planning for Pets in a Fallout Scenario

Sustainable Supplies:

Food and Water: Plan for long-term sustainability by storing enough food and water for your pets to last several weeks or even months. Rotate supplies regularly to ensure freshness.

Growing Pet Food: If possible, consider growing some of your pet's food, such as herbs for small animals or vegetables for certain pets. This can help supplement their diet if commercial food becomes scarce.

Backup Caretakers:

Designate Caretakers: Identify someone who can care for your pets if you are unable to. Ensure they are familiar with your pets' needs and have access to your emergency supplies.

Communication Plans: Establish a communication plan with your designated caretaker so they can be quickly informed if they need to step in.

Financial Considerations:

Pet Insurance: Consider pet insurance that covers emergencies and illnesses, including those related to radiation exposure. This can help cover veterinary costs in the aftermath of a fallout event.

Savings Fund: Set aside a savings fund specifically for pet-related emergencies, including medical care and additional supplies.

Protecting your pets in a fallout scenario requires careful planning and preparedness. By assembling an emergency kit, creating a safe sheltering environment, and understanding how to care for your pets during and after the fallout, you can ensure their safety and well-being. Remember, your pets rely on you for protection, and by being proactive, you can minimize their exposure to radiation and keep them safe during a nuclear emergency. With the right preparation, you and your pets can navigate a fallout scenario together, emerging healthy and secure on the other side.

Case Studies: Successful Fallout Shelter Experiences

Case studies of successful fallout shelter experiences provide valuable insights into the practical application of preparedness strategies and the effectiveness of fallout shelters during nuclear events. These real-world examples highlight the importance of proper planning, the adaptability of individuals and communities, and the lessons learned from those who have faced the challenges of sheltering during a fallout scenario. This chapter will examine several case studies, focusing on the key factors that contributed to successful outcomes.

The Greenbrier Bunker: A Secret Government Shelter

Background:

Location: The Greenbrier Bunker, also known as "Project Greek Island," was a massive fallout shelter constructed beneath The Greenbrier Resort in West Virginia during the Cold War. It was designed to house the United States Congress in the event of a nuclear attack.

Purpose: The bunker was intended to ensure the continuity of government by providing a secure location where members of Congress could continue to operate during a nuclear emergency.

Key Features:

Design and Construction: The Greenbrier Bunker was built to withstand a nuclear blast and protect against radiation. It included dormitories, meeting rooms, a medical clinic, and a broadcast center. The facility was equipped with air filtration systems, water purification, and food storage to support up to 1,000 people for several weeks.

Secrecy and Maintenance: The existence of the bunker was kept secret for over 30 years, and it was maintained in a state of readiness by a team of workers who posed as hotel staff. The facility was regularly updated to keep pace with technological advancements.

Outcome:

Decommissioning: The Greenbrier Bunker was never used for its intended purpose and was decommissioned in 1992 after its existence was revealed by a journalist. However, its construction and maintenance demonstrated the feasibility of large-scale fallout shelters and the importance of planning for continuity of government.

Legacy: The bunker remains a symbol of Cold War-era preparedness and is now a museum, offering tours to the public. It serves as an example of how extensive planning and resources can create a highly effective fallout shelter, even if never utilized.

The Swiss Civil Defense System: A National Approach to Fallout Sheltering

Background:

National Program: Switzerland is renowned for its comprehensive civil defense system, which includes a network of fallout shelters capable of accommodating the entire population. This program was developed in response to the threat of nuclear war during the Cold War.

Legal Mandate: Swiss law requires that all residential buildings include a fallout shelter, and municipalities are responsible for ensuring that shelters are available for all residents. The Swiss government also invested in public education and regular drills to prepare the population.

Key Features:

Widespread Shelters: Swiss fallout shelters are integrated into both private homes and public buildings. These shelters are designed to protect against radiation and provide for basic needs, including air filtration, water, food, and sanitation.

Community Involvement: Swiss citizens are educated about the importance of fallout shelters and are trained in their use. The government regularly conducts drills and updates the public on civil defense measures.

Outcome:

High Readiness: The Swiss civil defense system is considered one of the most effective in the world. The widespread availability of shelters, combined with public education and regular drills, ensures that the population is well-prepared for a nuclear event.

Model for Other Nations: Switzerland's approach to fallout sheltering has been studied by other countries as a model of effective civil defense. The program's success demonstrates the value of national-level planning and the importance of integrating shelters into everyday life.

The Cuban Missile Crisis: Improvised Sheltering in the United States

Background:

Crisis Overview: During the Cuban Missile Crisis in October 1962, the United States faced the imminent threat of nuclear war with the Soviet Union. This led to widespread fear and an urgent need for fallout protection.

Public Response: In response to the crisis, many Americans took steps to protect themselves, including improvising fallout shelters in basements, storm cellars, and other locations.

Key Features:

Improvised Shelters: With limited time and resources, many Americans used available materials to create makeshift fallout shelters. These often included reinforced basements with extra layers of protection, such as sandbags, concrete blocks, and lead sheeting to reduce radiation exposure.

Government Guidance: The U.S. government provided guidance on how to build and stock fallout shelters, including distributing pamphlets and broadcasting instructions on radio and television. Some communities also identified public buildings that could serve as shelters.

Outcome:

Avoidance of Nuclear Conflict: The Cuban Missile Crisis was ultimately resolved through diplomatic negotiations, and the improvised shelters were not tested in a real nuclear event. However, the crisis underscored the importance of preparedness and the ability of individuals to take protective measures under short notice.

Public Awareness: The crisis heightened public awareness of nuclear threats and led to increased interest in civil defense. Many of the shelters built during this time remained in place for years, serving as a reminder of the importance of preparedness.

The Fukushima Daiichi Nuclear Disaster: Sheltering and Evacuation in Japan

Background:

Disaster Overview: On March 11, 2011, a massive earthquake and tsunami struck Japan, leading to the failure of cooling systems at the Fukushima Daiichi Nuclear Power Plant. This resulted in the release of radioactive material and a widespread evacuation.

Government Response: The Japanese government issued evacuation orders for residents living within a 20-kilometer radius of the plant, while those living further away were advised to shelter in place to reduce radiation exposure.

Key Features:

Sheltering in Place: Residents who were unable to evacuate immediately were instructed to stay indoors, close windows and doors, and turn off ventilation systems to minimize exposure to radioactive particles. The government provided guidance on how to safely shelter in place and later distributed iodine tablets to protect against thyroid cancer.

Evacuation Logistics: The evacuation of over 160,000 people was a massive logistical challenge, complicated by the damage from the earthquake and tsunami. Evacuees were relocated to temporary shelters, where they received medical care and radiation monitoring.

Outcome:

Effective Sheltering: The shelter-in-place orders helped reduce radiation exposure for those who could not immediately evacuate. The distribution of iodine tablets and other protective measures also mitigated the health impact of radiation.

Challenges and Lessons: The Fukushima disaster highlighted the importance of timely and clear communication during a nuclear emergency. It also underscored the need for robust evacuation plans that account for multiple hazards, such as natural disasters and radiation.

The Community Fallout Shelter in Kansas: A Grassroots Effort

Background:

Community Initiative: In the 1960s, a small community in Kansas took the initiative to build a fallout shelter that could accommodate several families. The shelter was constructed with local materials and volunteer labor, reflecting a strong sense of community and preparedness.

Shelter Design: The shelter was built underground, with reinforced concrete walls and a heavy steel door. It was equipped with a hand-crank air filtration system, water storage, and food supplies to sustain the occupants for up to a month.

Key Features:

Collaborative Effort: The project was a collaborative effort, with community members pooling resources and skills to build the shelter. This grassroots approach fostered a strong sense of ownership and responsibility among the participants.

Preparedness Education: The community also engaged in regular drills and preparedness education, ensuring that all members knew how to use the shelter and understood the principles of radiation protection.

Outcome:

Community Resilience: Although the shelter was never used in a real nuclear event, it served as a powerful example of community resilience and the value of collective action in preparing for emergencies.

Legacy of Preparedness: The experience reinforced the importance of local initiatives in disaster preparedness. It demonstrated that even small communities can effectively protect themselves with the right planning and cooperation.

These case studies illustrate a range of successful fallout shelter experiences, from large-scale government projects to grassroots community efforts. The common thread across these examples is the importance of planning, collaboration, and adaptability in the face of potential nuclear threats.

Whether through government-led initiatives, community action, or individual preparedness, these case studies highlight the critical role that shelters can play in protecting lives during a nuclear event. They also provide valuable lessons that can be applied to future efforts to enhance nuclear preparedness and resilience. By learning from these successful experiences, we can better understand the key factors that contribute to effective fallout sheltering and apply these lessons to improve our own preparedness plans.

Common Myths and Misconceptions about Fallout Protection

Understanding the facts about fallout protection is essential for effective preparedness. However, many myths and misconceptions surround the topic, leading to confusion and potentially dangerous misunderstandings. This chapter will address some of the most common myths and misconceptions about fallout protection, providing clear explanations to help you make informed decisions about how to protect yourself and your loved ones during a nuclear event.

Myth: Fallout Shelters Are Only for the Wealthy

Reality:

Accessible Solutions: One of the most persistent myths is that fallout shelters are expensive and only accessible to the wealthy. While high-end shelters with luxury features do exist, effective fallout protection does not require a massive budget. Many practical, low-cost options are available, including using existing spaces in your home, such as basements or interior rooms, with some modifications to enhance their protective capabilities.

DIY Shelters: With basic construction skills and materials, you can create a fallout shelter that provides adequate protection against radiation. Sandbags, concrete blocks, and other inexpensive materials can be used to reinforce walls and reduce radiation exposure. The key is to focus on practical solutions that maximize protection within your budget.

Myth: You Need to Stay in a Fallout Shelter for Years

Reality:

Sheltering Duration: The idea that you need to remain in a fallout shelter for years is a common misconception. In reality, the most dangerous period following a nuclear event is the first few days to weeks when radiation levels are highest. Typically, you may need to stay in a well-shielded shelter for about 48 hours to two weeks, depending on the severity of the fallout and your location relative to the blast.

Gradual Re-entry: After the initial period, radiation levels will decrease significantly, allowing for short, controlled trips outside the shelter for essential tasks. However, it's important to continue monitoring radiation levels and follow official guidance before fully re-entering the outside environment.

Myth: Fallout Shelters Are Indestructible

Reality:

Limitations of Shelters: While fallout shelters are designed to protect against radiation, they are not indestructible. A direct hit from a nuclear blast can destroy most structures, including shelters. Fallout shelters are most effective when located at a safe distance from the blast site, where they can shield against radiation rather than withstand the physical force of the explosion.

Proper Location: The effectiveness of a fallout shelter largely depends on its location, design, and the materials used in its construction. It's crucial to have a realistic understanding of what your shelter can and cannot do, focusing on radiation protection rather than blast resistance.

Myth: Ordinary Clothing Can Protect You from Radiation

Reality:

Limited Protection: Ordinary clothing offers minimal protection against radiation. While clothing can help reduce the amount of radioactive particles that come into direct contact with your skin, it does not shield you from penetrating radiation like gamma rays.

Protective Gear: Specialized protective clothing, such as lead-lined garments or suits made from materials that block radiation, provides better protection but is typically used by professionals in controlled environments. For most people, the best protection is provided by staying indoors in a properly shielded shelter.

Myth: Once Fallout Settles, It's Safe to Go Outside

Reality:

Lingering Radiation: Fallout consists of radioactive particles that can remain hazardous for days, weeks, or even longer after they settle. The initial fallout period is the most dangerous, but lingering radiation can still pose serious health risks. Going outside too soon, even after fallout has settled, can result in exposure to harmful levels of radiation.

Monitoring and Caution: It's important to use radiation detection equipment to monitor levels before leaving the shelter. Only when radiation levels have decreased to safe levels, as indicated by reliable measurements, should you consider re-entering the outside environment.

Myth: Radiation Sickness Is Always Immediate

Reality:

Delayed Symptoms: Radiation sickness can develop over time, and symptoms may not appear immediately after exposure. Early symptoms of radiation sickness, such as nausea, vomiting, and fatigue, may occur within hours, but more severe effects, such as hair loss, skin burns, and internal organ damage, can take days or even weeks to manifest.

Long-Term Health Effects: Long-term exposure to lower levels of radiation can increase the risk of cancer and other health problems, which might not become apparent until years after the exposure. Understanding the potential for delayed effects underscores the importance of taking protective measures even if you don't feel sick right away.

Myth: If You Can't Build a Professional Fallout Shelter, There's No Point in Trying

Reality:

Improvised Shelters: Even if you don't have a professional fallout shelter, you can still create a safe space that significantly reduces radiation exposure. Improvised shelters, such as a basement with reinforced walls or an interior room with additional shielding, can provide substantial protection.

Practical Measures: The key is to focus on practical measures you can take with the resources available to you. Adding layers of dense materials like concrete, earth, or even books and furniture can help shield against radiation. The goal is to reduce radiation exposure as much as possible, even if you don't have a fully equipped shelter.

Myth: Fallout Only Affects the Area near the Blast

Reality:

Wide Area of Impact: Fallout can affect areas far beyond the immediate blast zone. Radioactive particles can be carried by wind and weather patterns, spreading over hundreds or even thousands of miles. This means that areas far from the blast site can still experience dangerous levels of radiation.

Importance of Monitoring: It's crucial to stay informed about the direction and extent of fallout spread, which can be influenced by factors such as wind speed and precipitation. Rely on official advisories and use radiation monitoring equipment to assess the safety of your location.

Myth: Drinking Water and Food Are Always Contaminated After Fallout

Reality:

Varying Contamination Levels: Not all water and food will be contaminated after a fallout event. Water from underground sources, such as wells, is less likely to be contaminated than surface water. Similarly, food stored in sealed containers or inside buildings is generally safer than food exposed to the open air.

Purification Methods: Contaminated water can often be purified through filtration, boiling, or the use of purification tablets. It's important to assess the level of contamination before consuming water or food and to use appropriate purification methods when necessary.

Myth: You Can't Survive a Nuclear Attack

Reality:

Survivability: With proper planning, preparation, and sheltering, it is possible to survive a nuclear attack, particularly if you are outside the immediate blast zone. The key to survival is reducing exposure to radiation, securing a safe shelter, and following emergency protocols.

Preparedness Saves Lives: Historical examples and studies show that people who take appropriate protective measures, such as seeking shelter and following official guidance, have a significantly higher chance of surviving and minimizing the effects of a nuclear event.

Myths and misconceptions about fallout protection can lead to dangerous decisions during a nuclear event. By understanding the facts and debunking these myths, you can take the necessary steps to protect yourself and your loved ones effectively. Remember, while nuclear events are terrifying, informed preparedness can make a significant difference in your ability to survive and recover. Focus on practical, science-based strategies for fallout protection, and don't let misconceptions prevent you from taking action that could save lives.

Legal Considerations: Zoning, Permits, and Property Rights

When planning to build or modify a fallout shelter, it's essential to consider the legal aspects of your project. Zoning laws, building permits, and property rights can significantly impact your ability to construct a shelter, especially if you live in an area with strict regulations. Understanding these legal considerations can help you avoid potential issues and ensure that your fallout shelter is both safe and compliant with local laws. This chapter will cover the key legal considerations, including zoning regulations, permits, property rights, and potential legal challenges you may encounter.

Understanding Zoning Laws

What Are Zoning Laws?:

Definition and Purpose: Zoning laws are local regulations that govern how land can be used in specific areas. These laws are designed to control land use in a way that promotes orderly development, protects property values, and ensures public safety. Zoning laws can dictate what types of structures can be built, where they can be located, and how they can be used.

Impact on Fallout Shelters: Depending on your location, zoning laws may restrict the construction of underground structures, such as fallout shelters, or impose specific requirements on their design and placement. For example, certain residential zones may prohibit or limit the construction of underground bunkers due to concerns about land stability, drainage, or potential impact on neighboring properties.

Researching Local Zoning Regulations:

Consulting Zoning Maps: Zoning maps are available through your local government's planning or zoning office. These maps detail the zoning classifications for different areas and provide information on the types of structures allowed in each zone.

Zoning Codes and Ordinances: Review the zoning codes and ordinances for your area to understand any restrictions or requirements related to underground construction. Zoning codes will outline what is permissible in your zone and any conditions that must be met.

Seeking Variances and Exceptions:

Applying for a Variance: If your zoning laws do not permit the construction of a fallout shelter, you may be able to apply for a variance. A variance is a legal exception that allows you to deviate from the zoning requirements under certain conditions. To obtain a variance, you will typically need to demonstrate that the shelter will not negatively impact the surrounding area and that the variance is necessary for your safety or property use.

Presenting Your Case: When applying for a variance, be prepared to present your case to the local zoning board or planning commission. You may need to provide detailed plans, engineering reports, and evidence that your shelter will not harm the environment, neighboring properties, or the community.

Obtaining Building Permits

The Importance of Permits:

Ensuring Compliance: Building permits are required for most construction projects, including fallout shelters, to ensure that the work complies with local building codes and safety standards. Obtaining a permit is essential to avoid legal issues and ensure that your shelter is built to a safe standard.

Inspection Process: When you apply for a building permit, your local building department will review your plans to ensure they meet the required codes. After construction begins, inspectors may visit your site to verify that the work is being carried out according to the approved plans.

Steps to Obtain a Permit:

Submitting an Application: The first step in obtaining a building permit is to submit an application to your local building department. Your application should include detailed plans for your shelter, including dimensions, materials, structural design, and any additional features like ventilation or plumbing systems.

Providing Documentation: Along with your application, you may need to provide additional documentation, such as soil surveys, engineering reports, and proof of property ownership. Some jurisdictions may also require environmental impact assessments, especially if your project could affect local water tables or wildlife habitats.

Paying Fees: Building permit fees vary depending on the scope and complexity of your project. Be prepared to pay these fees as part of the application process. These fees cover the cost of plan review and inspections.

Navigating the Approval Process:

Review and Approval: After you submit your application, the building department will review your plans. This review process may take several weeks, depending on the complexity of your project and the workload of the department.

Addressing Concerns: If the building department identifies any issues with your plans, they may ask you to make revisions before granting the permit. Be prepared to work with your contractor or engineer to address these concerns and resubmit the revised plans.

Final Inspection and Approval:

Inspection Stages: Once construction begins, your project will undergo several inspections at different stages. These inspections ensure that the work meets the approved plans and complies with building codes.

Final Approval: After the final inspection, if everything meets the required standards, you will receive a certificate of occupancy or final approval for your shelter. This document confirms that your shelter is safe and legal to use.

Property Rights and Legal Considerations

Understanding Property Rights:

Ownership Rights: Property rights refer to the legal rights you have over the land you own, including the right to build structures like fallout shelters. However, these rights can be limited by zoning laws, easements, and other legal restrictions.

Easements and Restrictions: Easements are legal rights granted to others to use part of your property for specific purposes, such as utility lines or access roads. Before building a shelter, check for any easements or deed restrictions on your property that could limit where or how you can build.

Dealing with Neighboring Properties:

Impact on Neighbors: When constructing a fallout shelter, consider how it might impact neighboring properties. For example, excavation work could affect land stability or drainage patterns, leading to potential disputes with neighbors.

Boundary Lines: Ensure that your shelter is built within the boundaries of your property. Encroaching on a neighbor's land, even accidentally, can lead to legal disputes and potential court action.

Liability and Insurance:

Liability Considerations: As a property owner, you are responsible for ensuring that your construction project does not pose a hazard to others. If someone is injured due to your shelter construction, you could be held liable. Ensuring compliance with building codes and safety standards can help mitigate this risk.

Insurance Coverage: Check with your insurance provider to ensure that your fallout shelter is covered under your homeowner's policy. You may need to update your policy to include the shelter or obtain additional coverage for construction-related risks.

Legal Challenges and Dispute Resolution

Common Legal Challenges:

Neighbor Disputes: Disputes with neighbors can arise over issues such as noise, land use, or perceived safety risks associated with your fallout shelter. It's important to address these concerns proactively and seek resolution through dialogue or mediation before they escalate into legal battles.

Zoning and Permit Denials: If your zoning variance or building permit application is denied, you have the right to appeal the decision. This process typically involves presenting your case to a higher authority, such as a zoning board of appeals or a local court.

Resolving Disputes:

Mediation and Negotiation: Many legal disputes can be resolved through mediation or negotiation, avoiding the need for costly and time-consuming litigation. Consider hiring a mediator or engaging in direct discussions with the involved parties to find a mutually agreeable solution.

Legal Representation: If you face significant legal challenges, such as a lawsuit or a complex zoning issue, it may be wise to consult with an attorney who specializes in real estate or construction law. Legal representation can help you navigate the complexities of the legal system and protect your interests.

Special Considerations for Community Fallout Shelters

Community Projects:

Collaborative Efforts: Building a community fallout shelter involves additional legal considerations, such as shared ownership, liability, and governance. It's important to establish clear agreements among all participants regarding the construction, maintenance, and use of the shelter.

Legal Agreements: Consider drafting a formal agreement or contract that outlines the responsibilities of each participant, how costs will be shared, and how decisions will be made. This agreement should be reviewed by a legal professional to ensure it is legally binding and fair to all parties.

Public Shelters and Municipal Regulations:

Municipal Shelters: In some cases, local governments may support the construction of public fallout shelters. These shelters may be subject to different regulations and permit requirements than private shelters. Coordination with local authorities is essential to ensure compliance.

Public Access and Liability: Public fallout shelters must comply with accessibility standards, such as the Americans with Disabilities Act (ADA) in the United States. Additionally, the managing entity must consider liability issues and insurance coverage to protect against potential legal claims.

Navigating the legal landscape of fallout shelter construction requires careful attention to zoning laws, building permits, property rights, and potential legal challenges. By understanding these considerations and taking proactive steps to comply with local regulations, you can ensure that your fallout shelter is not only safe and functional but also legally sound. Whether you're building a private shelter for your family or participating in a community project, it's important to approach the process with thorough planning and a clear understanding of the legal requirements. With the right preparation and legal knowledge, you can protect yourself and your property while contributing to your overall preparedness for a nuclear event.

Insurance and Fallout Shelters: What You Should Know

When constructing or maintaining a fallout shelter, understanding the role of insurance is crucial. Insurance can provide financial protection for your investment in the shelter and cover potential risks associated with its use or construction. However, navigating the complexities of insurance policies in relation to fallout shelters requires careful consideration of coverage options, exclusions, and additional endorsements. This chapter will explore what you need to know about insuring a fallout shelter, including coverage types, factors that influence premiums, potential exclusions, and tips for securing the best insurance for your needs.

Homeowners Insurance and Fallout Shelters

Standard Homeowners Insurance Policies:

Coverage for Structures: In most cases, standard homeowners insurance policies cover structures on your property, including fallout shelters, as part of the dwelling or other structures coverage. This typically includes protection against risks such as fire, wind damage, and vandalism. However, coverage for fallout shelters may not be explicitly stated, so it's important to clarify this with your insurance provider.

Personal Property: Personal property stored in your fallout shelter, such as emergency supplies or valuable items, may also be covered under your homeowners policy. However, coverage limits and exclusions apply, so review your policy to understand what is and isn't covered.

Building a New Fallout Shelter:

Construction Phase Coverage: If you are constructing a new fallout shelter, ensure that your homeowners insurance provides coverage during the construction phase. This coverage is essential in case of accidents, theft, or damage to materials during the build. You may need to notify your insurer of the construction project and potentially add a builder's risk policy or endorsement to cover these risks.

Property Value and Coverage Adjustments: After the shelter is completed, your property value may increase, which could affect your insurance coverage. Notify your insurance provider of the new structure to ensure that your policy limits are adequate to cover the full value of your property, including the shelter.

Specialized Coverage for Fallout Shelters

Endorsements and Riders:

Structural Endorsements: Some insurance companies offer endorsements or riders specifically designed for fallout shelters or underground structures. These endorsements provide additional coverage beyond what is included in a standard homeowners policy, such as protection against specific risks associated with underground construction.

Earthquake and Flood Insurance: Depending on your location, you may need additional coverage for risks not typically included in standard policies, such as earthquakes or floods. Since fallout shelters are often underground, they may be more susceptible to damage from these events. Consider adding earthquake or flood insurance to protect your shelter.

Liability Insurance:

Increased Liability Risks: Owning a fallout shelter may increase your liability risks, especially if others are using the shelter or if it is part of a community project. Liability coverage protects you in case someone is injured on your property or if your shelter causes damage to neighboring properties.

Umbrella Policies: If you anticipate higher liability risks, consider purchasing an umbrella policy, which provides additional liability coverage beyond the limits of your standard homeowners policy. This can offer extra protection in case of lawsuits or significant claims.

Factors Influencing Insurance Premiums

Construction Materials and Design:

Material Durability: The materials used in the construction of your fallout shelter can impact your insurance premiums. Shelters made from durable, fire-resistant materials such as reinforced concrete may be viewed as less risky by insurers, potentially leading to lower premiums.

Safety Features: The presence of safety features, such as proper ventilation, air filtration systems, and secure entrances, can also influence your premiums. These features demonstrate that the shelter is designed to protect occupants and reduce risks, which can be favorable to insurers.

Location and Environmental Risks:

Geographical Location: The location of your property and the associated environmental risks, such as proximity to fault lines, flood zones, or high-wind areas, will affect your insurance rates. Properties in high-risk areas may require additional coverage or result in higher premiums.

Proximity to Nuclear Facilities: If your property is near a nuclear power plant or other potential nuclear hazards, your insurance premiums may be influenced by the perceived risk of a nuclear event. However, this is typically a minor factor compared to other environmental risks.

Usage of the Shelter:

Primary vs. Secondary Use: If your fallout shelter is intended for primary use (such as a living space or home office), it may be considered differently than if it is solely for emergency use. Regular use of the shelter may require additional coverage and could impact your premiums.

Community Shelters: If your fallout shelter is part of a community effort and will be used by multiple households, this could affect your insurance coverage. Ensure that all participants are adequately covered and that liability is clearly defined in legal agreements.

Exclusions and Limitations in Fallout Shelter Insurance

Common Exclusions:

Nuclear Exclusions: Many standard insurance policies include a nuclear exclusion clause, which means that damages resulting directly from a nuclear explosion or radiation are not covered. This is a standard exclusion in most policies and reflects the catastrophic nature of nuclear events.

Wear and Tear: Standard policies typically do not cover damage resulting from normal wear and tear, deterioration, or lack of maintenance. This means that if your fallout shelter suffers from issues like mold, structural decay, or other maintenance-related problems, your insurance may not cover the repairs.

Understanding Policy Limitations:

Coverage Limits: Be aware of the coverage limits on your policy, especially for additional structures like fallout shelters. If the value of your shelter exceeds the standard coverage limit, you may need to purchase additional coverage or increase your policy limits.

Claiming for Fallout-Related Damage: If your fallout shelter is damaged by an event that is not explicitly excluded, such as an earthquake or fire, you should be able to file a claim. However, you must provide documentation proving the cause of the damage and that it falls within the scope of your coverage.

Tips for Securing Fallout Shelter Insurance

Reviewing and Updating Your Policy:

Regular Policy Reviews: Review your insurance policy regularly, especially after completing the construction of a new shelter or making significant upgrades. Ensure that your coverage reflects the current value of your property and shelter.

Comparing Quotes: Shop around for insurance quotes from different providers, especially if you need specialized coverage for a fallout shelter. Comparing quotes can help you find the best coverage at a competitive price.

Documenting Your Shelter:

Detailed Documentation: Maintain detailed records of your fallout shelter, including construction plans, materials used, safety features, and any upgrades. Photographs and videos of the shelter can also be helpful when filing a claim.

Keeping Receipts and Records: Keep all receipts for materials, labor, and any repairs or maintenance work performed on your shelter. This documentation can be crucial in proving the value of the shelter and justifying claims in the event of damage.

Consulting with an Insurance Professional:

Expert Advice: Consider consulting with an insurance professional who has experience with fallout shelters or similar structures. They can help you navigate the complexities of insurance, recommend the right coverage, and identify potential gaps in your policy.

Customized Coverage: An insurance professional can also help you customize your coverage based on your specific needs and the unique features of your fallout shelter, ensuring that you are adequately protected.

Insuring a fallout shelter requires careful consideration of various factors, including the type of coverage needed, the risks associated with your location, and the specific features of your shelter. By understanding these factors and working closely with your insurance provider, you can secure the protection you need to safeguard your investment and ensure peace of mind. Whether you're constructing a new shelter or maintaining an existing one, regular policy reviews, thorough documentation, and expert guidance can help you navigate the complexities of fallout shelter insurance. With the right coverage in place, you can be confident that your shelter is protected against a range of risks, allowing you to focus on preparedness and safety in the event of a nuclear emergency.

The Role of Civil Defense in Nuclear Fallout Scenarios

Civil defense plays a crucial role in protecting the public during nuclear fallout scenarios. Historically, civil defense programs have been established by governments to prepare citizens for the potential impact of nuclear warfare, including providing information, training, resources, and infrastructure to mitigate the effects of a nuclear event. Understanding the role of civil defense in nuclear fallout scenarios helps individuals and communities better prepare for such emergencies and coordinate with official response efforts. This chapter will explore the various aspects of civil defense, including its history, current practices, key components, and how individuals can engage with and benefit from these programs.

The History of Civil Defense

Origins and Development:

World War II and Early Cold War: The concept of civil defense emerged during World War II and expanded during the early Cold War as nations recognized the potential for large-scale destruction from nuclear weapons. Governments began to establish formal civil defense programs aimed at protecting civilians from the effects of air raids and, later, nuclear attacks.

Public Education and Drills: Early civil defense efforts focused on educating the public about the dangers of nuclear weapons and how to protect themselves. This included distributing pamphlets, broadcasting public service announcements, and conducting air raid and "duck and cover" drills in schools and communities.

Fallout Shelters: As the threat of nuclear war grew, governments promoted the construction of fallout shelters, both public and private, as a key component of civil defense. Some countries, like Switzerland, developed extensive shelter networks, while others, like the United States, encouraged citizens to build shelters in their homes.

Cold War Civil Defense Programs:

U.S. Civil Defense: In the United States, the Federal Civil Defense Administration (FCDA) was established in 1950 to oversee civil defense efforts, including shelter construction, evacuation planning, and public education. The program evolved over the decades, with varying levels of public interest and government funding.

Global Efforts: Other countries, particularly those in Europe, also developed robust civil defense programs. For example, Sweden and Switzerland built extensive shelter systems capable of protecting large portions of their populations, while the United Kingdom focused on public education and emergency planning.

Decline and Modernization:

Post-Cold War Shift: With the end of the Cold War, the perceived threat of nuclear war diminished, leading to a decline in civil defense programs in many countries. However, the rise of new threats, such as terrorism and natural disasters, led to the integration of civil defense into broader emergency management frameworks.

Modern Civil Defense: Today, civil defense efforts are often part of national emergency management agencies, focusing on all-hazards preparedness, including nuclear events. These programs emphasize resilience, continuity of government, and public safety in the face of various threats.

Key Components of Civil Defense in Nuclear Scenarios

Public Education and Awareness:

Information Dissemination: A critical component of civil defense is providing the public with accurate, timely information about the risks of nuclear fallout and the steps they can take to protect themselves. This includes guidance on sheltering, evacuation, and decontamination procedures.

Preparedness Campaigns: Governments may launch preparedness campaigns that encourage citizens to develop emergency plans, stockpile supplies, and participate in drills. These campaigns often use multiple channels, including television, radio, social media, and printed materials, to reach a broad audience.

Sheltering and Infrastructure:

Public Shelters: Civil defense programs may include the construction and maintenance of public fallout shelters, particularly in urban areas. These shelters are designed to protect large numbers of people from radiation and provide basic necessities, such as food, water, and medical care.

Private Shelter Support: Some civil defense initiatives offer guidance or incentives for individuals to build private fallout shelters. This support may include technical advice, subsidies, or tax incentives to encourage shelter construction.

Emergency Response Coordination:

Evacuation Planning: Civil defense programs often include detailed evacuation plans for areas at high risk of nuclear fallout. These plans involve coordinated efforts between local, regional, and national authorities to safely move people out of danger zones.

First Responders and Training: Civil defense also involves training first responders, such as firefighters, police officers, and medical personnel, to handle nuclear emergencies. This training includes radiation detection, decontamination procedures, and providing medical care for radiation exposure.

Communication Systems:

Emergency Alert Systems: Effective communication is vital in a nuclear fallout scenario. Civil defense programs typically establish emergency alert systems that can quickly disseminate warnings and instructions to the public through various channels, including sirens, radio broadcasts, and mobile alerts.

Public Access to Information: In addition to real-time alerts, civil defense programs may provide access to information about radiation levels, safe zones, and available shelters through websites, hotlines, and mobile apps.

International Collaboration:

Global Coordination: In a nuclear fallout scenario, international collaboration is essential, especially in regions where fallout can cross borders. Civil defense agencies often work with international organizations, such as the International Atomic Energy Agency (IAEA), to share information, resources, and best practices.

Mutual Aid Agreements: Countries may enter into mutual aid agreements that facilitate the sharing of resources, personnel, and expertise during a nuclear emergency. These agreements can enhance the overall effectiveness of civil defense efforts and ensure a coordinated response.

How Individuals Can Engage with Civil Defense Programs

Participating in Drills and Training:

Community Drills: Many civil defense programs conduct regular community drills that simulate nuclear fallout scenarios. These drills provide an opportunity for individuals and families to practice emergency procedures, such as sheltering and evacuation, in a controlled environment.

CERT Programs: The Community Emergency Response Team (CERT) program is one example of how individuals can get involved in civil defense. CERT programs train volunteers in basic disaster response skills, including nuclear safety, and encourage community involvement in preparedness efforts.

Staying Informed:

Signing Up for Alerts: Individuals can stay informed about nuclear threats by signing up for emergency alerts from their local or national civil defense agency. These alerts provide real-time information about potential threats and recommended actions.

Educational Resources: Civil defense agencies often provide educational resources, such as brochures, websites, and instructional videos, to help individuals understand the risks of nuclear fallout and how to prepare. Taking advantage of these resources can enhance personal and family preparedness.

Building a Personal Emergency Plan:

Creating a Plan: Developing a personal or family emergency plan that aligns with civil defense guidelines is essential. This plan should include details about where to shelter, how to communicate with family members, and what supplies to stockpile.

Practicing the Plan: Regularly practicing your emergency plan with your family ensures that everyone knows what to do in a nuclear fallout scenario. Familiarity with the plan can reduce panic and increase the likelihood of a successful response.

The Future of Civil Defense

Adapting to New Threats:

Evolving Risks: As global security dynamics change, civil defense programs must adapt to new and emerging threats, such as cyber-attacks on nuclear facilities or the proliferation of small-scale nuclear weapons. Modern civil defense efforts increasingly focus on flexibility and resilience to address a wide range of potential scenarios.

Technological Advancements: Advances in technology, such as improved radiation detection, enhanced communication systems, and better predictive modeling, are likely to shape the future of civil defense. These technologies can improve the accuracy and speed of threat assessments and responses.

Strengthening Community Resilience:

Local Preparedness Initiatives: Encouraging local communities to take an active role in civil defense can enhance overall resilience. This includes promoting neighborhood-based preparedness groups, local shelter construction, and community education programs.

Public-Private Partnerships: Collaborations between government agencies, private sector companies, and non-governmental organizations (NGOs) can expand the reach and effectiveness of civil defense programs. These partnerships can help fund shelter construction, improve emergency response capabilities, and increase public awareness.

Global Civil Defense Networks:

International Cooperation: Strengthening international civil defense networks is crucial for addressing cross-border nuclear threats and sharing best practices. By participating in global initiatives and exercises, countries can improve their preparedness and response capabilities.

Standardization of Practices: Developing standardized guidelines and protocols for nuclear fallout scenarios can ensure a more consistent and effective global response. International organizations can play a key role in coordinating these efforts and promoting uniform civil defense standards.

Civil defense remains a vital component of national security and public safety in the face of nuclear fallout scenarios. By understanding the role of civil defense, individuals and communities can better prepare for nuclear emergencies and work in coordination with official response efforts. Whether through public education, participation in drills, or staying informed about threats, there are many ways individuals can engage with civil defense programs to enhance their preparedness.

Fallout and Climate Change: Analyzing the Connection

The connection between nuclear fallout and climate change is a complex and multifaceted issue that encompasses both the potential impact of nuclear events on the climate and the broader environmental consequences of nuclear warfare. While fallout from a single nuclear event primarily affects localized areas, large-scale nuclear conflicts, such as those involving multiple detonations, could have significant and far-reaching effects on the global climate. This chapter will explore the relationship between fallout and climate change, examining the potential environmental impacts, the concept of nuclear winter, and how these scenarios could interact with existing climate change trends.

The Environmental Impact of Nuclear Fallout

Radioactive Contamination:

Soil and Water Contamination: Fallout consists of radioactive particles that settle on the ground after a nuclear explosion. These particles can contaminate soil and water sources, leading to long-term environmental damage. Contaminated soil can become infertile, making agriculture impossible in affected areas, while contaminated water sources pose significant health risks to both humans and wildlife.

Bioaccumulation in Ecosystems: Radioactive materials can enter the food chain through bioaccumulation, where they are absorbed by plants and subsequently consumed by animals and humans. This can lead to widespread ecological disruption and long-term health effects in both animals and people who consume contaminated food.

Destruction of Ecosystems:

Immediate Effects: The immediate effects of a nuclear explosion, including intense heat, pressure waves, and radiation, can cause widespread destruction of ecosystems. Forests, wetlands, and other natural habitats can be obliterated, leading to the loss of biodiversity and the disruption of ecological balance.

Long-Term Recovery: Recovery from nuclear fallout can take decades or even centuries, depending on the severity of the contamination. In some cases, ecosystems may never fully recover, leading to permanent changes in the landscape and loss of species.

The Concept of Nuclear Winter

Nuclear Winter Theory:

Origins of the Theory: The concept of nuclear winter emerged in the 1980s, when scientists began studying the potential climate effects of a large-scale nuclear war. The theory suggests that multiple nuclear detonations could inject massive amounts of smoke, soot, and dust into the stratosphere, blocking sunlight and causing a significant drop in global temperatures.

Mechanism of Action: In a nuclear winter scenario, the soot and smoke from burning cities, forests, and other targets would rise into the upper atmosphere, where it could remain for months or even years. This would reduce the amount of sunlight reaching the Earth's surface, leading to a dramatic cooling effect, similar to that observed after large volcanic eruptions.

Potential Climate Effects:

Global Cooling: The primary effect of nuclear winter would be a significant and rapid cooling of the Earth's surface. This cooling could lead to shorter growing seasons, widespread crop failures, and famine. The extent of the cooling would depend on the scale of the nuclear conflict and the amount of soot injected into the atmosphere.

Disruption of Weather Patterns: A nuclear winter could also disrupt global weather patterns, leading to changes in precipitation, storm frequency, and wind patterns. These changes could exacerbate existing climate challenges, such as droughts, floods, and extreme weather events.

Impact on the Ozone Layer:

Ozone Depletion: In addition to causing cooling, the injection of smoke and soot into the atmosphere could lead to the depletion of the ozone layer. The ozone layer protects life on Earth from harmful ultraviolet (UV) radiation. A weakened ozone layer would result in higher levels of UV radiation reaching the surface, increasing the risk of skin cancer, cataracts, and other health issues, as well as damaging crops and marine ecosystems.

Interaction with Climate Change Trends

Climate Change and Environmental Vulnerability:

Increased Vulnerability: The ongoing effects of climate change, such as rising temperatures, melting ice caps, and changing precipitation patterns, have already made many ecosystems and communities more vulnerable to disruption. In this context, the impact of nuclear fallout or a nuclear winter could be even more devastating, as it would compound existing environmental stressors.

Feedback Loops: The interaction between nuclear fallout and climate change could create feedback loops that worsen both phenomena. For example, the cooling effect of a nuclear winter could temporarily mask global warming, but the eventual clearing of soot from the atmosphere could lead to a rapid rebound in temperatures, potentially exacerbating the effects of climate change.

Impact on Global Food Security:

Crop Failures: Both climate change and nuclear winter scenarios pose significant risks to global food security. Climate change is already leading to more frequent and severe droughts, floods, and heatwaves, which threaten agricultural productivity. The additional cooling and disruption of weather patterns caused by nuclear winter could lead to widespread crop failures, reducing food availability and increasing the risk of famine.

Distribution Challenges: In a nuclear fallout scenario, the destruction of infrastructure and the contamination of agricultural land could further complicate food distribution, making it difficult to transport and store food in affected regions. Combined with the effects of climate change, this could lead to severe food shortages and social instability.

Mitigation and Adaptation Strategies

Reducing the Risk of Nuclear Conflict:

Diplomacy and Disarmament: The most effective way to prevent the environmental and climate impacts of nuclear fallout is to reduce the risk of nuclear conflict through diplomacy, arms control, and disarmament.

International agreements, such as the Treaty on the Non-Proliferation of Nuclear Weapons (NPT) and the Comprehensive Nuclear-Test-Ban Treaty (CTBT), play a crucial role in limiting the spread of nuclear weapons and reducing the likelihood of their use.

Preventive Measures: Governments and international organizations must continue to pursue preventive measures, including dialogue between nuclear-armed states, confidence-building measures, and crisis management protocols, to reduce the risk of nuclear escalation.

Climate Change Mitigation and Adaptation:

Strengthening Resilience: To mitigate the potential combined effects of nuclear fallout and climate change, it is essential to strengthen the resilience of communities and ecosystems. This includes investing in sustainable agriculture, improving water management, and protecting biodiversity to ensure that natural systems can better withstand and recover from environmental shocks.

Preparedness Planning: Governments and communities should integrate nuclear fallout scenarios into broader climate adaptation and disaster preparedness plans. This includes developing strategies for food security, infrastructure protection, and public health in the face of multiple, overlapping environmental threats.

Public Awareness and Education:

Raising Awareness: Public awareness campaigns can help educate people about the potential connection between nuclear fallout and climate change, as well as the steps that can be taken to reduce these risks. Increasing public understanding of the potential consequences of nuclear conflict can also support efforts to promote disarmament and non-proliferation.

Community Engagement: Engaging communities in discussions about both climate change and nuclear risks can help build support for policies and actions that address these challenges. Local initiatives, such as community gardens, renewable energy projects, and disaster preparedness workshops, can empower individuals and groups to take action at the grassroots level.

The Global Responsibility to Address the Connection

International Cooperation:

Global Governance: Addressing the connection between nuclear fallout and climate change requires a coordinated global response. International organizations, such as the United Nations and the IAEA, must work together to promote policies that reduce the risk of nuclear conflict while also addressing the root causes of climate change.

Shared Responsibility: All nations have a role to play in preventing nuclear conflict and mitigating climate change. This includes committing to disarmament, reducing greenhouse gas emissions, and supporting international efforts to build a more sustainable and peaceful world.

Ethical Considerations:

Protecting Future Generations: The potential consequences of nuclear fallout and climate change pose significant ethical challenges. Protecting future generations from the devastating impacts of these threats requires bold action and a commitment to long-term sustainability. Governments, businesses, and individuals must all take responsibility for reducing these risks and ensuring a livable planet for future generations.

Equity and Justice: The impacts of both nuclear fallout and climate change are likely to be felt most acutely by vulnerable populations, including those in developing countries, marginalized communities, and future generations. Efforts to address these threats must be guided by principles of equity and justice, ensuring that the most vulnerable are protected and that the benefits of action are shared fairly.

The connection between nuclear fallout and climate change highlights the complex and interrelated challenges that the world faces in the 21st century. While the immediate effects of nuclear fallout are often localized, large-scale nuclear conflicts have the potential to cause significant global climate disruptions, compounding the existing challenges posed by climate change.

Addressing these risks requires a multifaceted approach that includes reducing the likelihood of nuclear conflict, strengthening resilience to environmental shocks, and promoting international cooperation. By understanding the potential consequences of nuclear fallout and climate change, and by taking proactive steps to mitigate these risks, we can work towards a safer, more sustainable future for all.

Ethical Considerations in Fallout Shelter Design

Designing fallout shelters involves not only technical and practical considerations but also significant ethical issues. These ethical considerations touch on questions of fairness, equity, access, and the potential societal impacts of sheltering during and after a nuclear event. As fallout shelters are designed to protect lives in extreme circumstances, the decisions made in their planning and construction can have far-reaching implications for individuals, communities, and society as a whole. This chapter will explore the ethical considerations in fallout shelter design, including issues of equity and access, the potential for social division, the responsibilities of shelter owners, and the broader societal implications.

Equity and Access to Fallout Shelters

Fair Access to Protection:

Socioeconomic Disparities: One of the primary ethical concerns in fallout shelter design is the disparity in access to protection based on socioeconomic status. Fallout shelters can be expensive to build, maintain, and equip, which may limit access to those with financial resources. This raises ethical questions about the fairness of a system where the wealthy can afford better protection in the event of a nuclear disaster, while those with fewer resources may be left vulnerable.

Public vs. Private Shelters: The existence of private fallout shelters, often accessible only to those who can afford them, contrasts with the need for public shelters that provide protection to all members of society, regardless of income. Ethical fallout shelter design should consider how to make protection more equitable, potentially through the development and maintenance of public shelters or government subsidies to support the construction of private shelters for those with limited means.

Government Responsibility:

Public Infrastructure: Governments have an ethical responsibility to protect their citizens, which may include providing access to public fallout shelters or ensuring that sufficient infrastructure exists to accommodate the population during a nuclear event. The design and distribution of these shelters should be guided by principles of fairness and inclusivity, ensuring that all citizens have an equal opportunity to access protection.

Subsidies and Incentives: To address disparities in access, governments could offer subsidies, grants, or tax incentives to help lower-income individuals and families build or retrofit shelters. This approach could help mitigate the ethical concerns related to socioeconomic disparities and promote broader access to protection.

The Potential for Social Division

Social Stratification:

Division by Wealth: The presence of high-end, luxurious fallout shelters that only the wealthy can afford has the potential to exacerbate social divisions. In a crisis situation, this could lead to increased tension and resentment between different socioeconomic groups, as those without access to shelters may perceive that their lives are valued less.

Exclusivity and Isolation: The use of private fallout shelters can also lead to social isolation, where individuals or families retreat into their own protective spaces, potentially fracturing community bonds. This exclusivity may hinder collective efforts to rebuild and recover after a nuclear event, as those who have sheltered privately may be less inclined to engage with broader community recovery efforts.

Community Cohesion:

Promoting Inclusivity: Ethical fallout shelter design should consider how to promote community cohesion rather than division. This could involve designing community shelters that accommodate multiple families or groups, fostering a sense of shared responsibility and mutual support. These shelters can serve as places where people come together during a crisis, strengthening community bonds and facilitating collective recovery efforts.

Addressing Stigmatization: There is also a risk that those who cannot afford private shelters or who must rely on public shelters may feel stigmatized or marginalized. Ethical considerations in shelter design should include efforts to minimize this stigmatization, perhaps by ensuring that public shelters are of high quality and provide similar levels of protection and comfort as private ones.

Responsibilities of Shelter Owners

Moral Obligations:

Offering Shelter to Others: Private shelter owners face ethical questions about their responsibilities to others in a crisis. Should a shelter owner be morally obligated to offer protection to neighbors or community members who do not have access to a shelter? This question is particularly pressing in densely populated areas where many people may be left without protection.

Capacity and Preparedness: Shelter owners should consider the ethical implications of their shelter's capacity. If a shelter is designed to accommodate more people than the immediate family, the owner may face difficult decisions about who to admit in an emergency. Ethical design should include considerations for accommodating as many people as possible, as well as clear guidelines for how to manage such situations.

Legal and Ethical Boundaries:

Legal Protections: In some jurisdictions, legal protections may exist for shelter owners, such as Good Samaritan laws that protect individuals from liability when offering aid during an emergency. However, the ethical responsibility of shelter owners may extend beyond legal requirements, particularly in life-and-death situations.

Ethical Dilemmas: Shelter owners may face ethical dilemmas, such as whether to prioritize family members, friends, or neighbors, and how to handle situations where the shelter's capacity is exceeded. Planning for these scenarios in advance, with clear ethical guidelines, can help mitigate the stress and moral ambiguity that may arise in a crisis.

Broader Societal Implications

Long-Term Social Impact:

Rebuilding Society: Fallout shelter design has implications for how society may rebuild after a nuclear event. Shelters that promote isolation and exclusivity may hinder efforts to rebuild a cohesive, functioning society. In contrast, shelters designed with community engagement in mind may facilitate more effective recovery and contribute to a more resilient social fabric.

Ethical Urban Planning: Urban planners and policymakers should consider the ethical implications of fallout shelters in the broader context of community design. Integrating shelters into public spaces, such as schools, community centers, and government buildings, can ensure more equitable access and promote social solidarity.

Ethical Design Principles:

Human Dignity: Fallout shelters should be designed with respect for human dignity, ensuring that all occupants have access to basic needs, such as food, water, sanitation, and medical care. The design should also consider the psychological well-being of occupants, providing spaces that are comfortable and conducive to mental health.

Sustainability and Environmental Impact: The environmental impact of fallout shelter construction and operation should be considered in the design process. Ethical design principles should prioritize sustainability, using materials and methods that minimize environmental harm and contribute to the long-term resilience of the community.

Addressing Ethical Challenges in Shelter Design

Inclusive Decision-Making:

Community Involvement: Engaging the community in the planning and design process for fallout shelters can help ensure that ethical considerations are addressed. This involvement can take the form of public consultations, focus groups, or collaborative design workshops, where community members have a voice in decisions that affect their safety and well-being.

Balancing Competing Interests: Ethical shelter design often involves balancing competing interests, such as individual rights, community needs, and public safety. Transparent decision-making processes that consider the perspectives of all stakeholders can help navigate these complex ethical issues.

Ethical Guidelines for Designers and Planners:

Codes of Ethics: Architects, engineers, and urban planners involved in fallout shelter design should adhere to professional codes of ethics that emphasize the importance of public safety, equity, and social responsibility. These codes can provide guidance on how to navigate the ethical challenges inherent in shelter design.

Ethical Training: Incorporating ethics training into the education and professional development of those involved in shelter design can help raise awareness of the ethical issues at play and encourage responsible decision-making.

The ethical considerations in fallout shelter design are multifaceted and deeply intertwined with questions of equity, access, and social responsibility. Designing shelters that protect human life while promoting fairness and community cohesion requires careful thought and a commitment to ethical principles.

By addressing these ethical considerations, shelter designers, owners, and policymakers can contribute to a more just and resilient society, where protection from nuclear threats is not just a privilege for the few but a right for all. As we continue to face the challenges of a complex and uncertain world, the ethical design of fallout shelters will play a crucial role in ensuring that we emerge from crises stronger and more united.

The Future of Fallout Protection: Innovations on the Horizon

As the world continues to evolve, so too does the technology and methodology behind fallout protection. The future of fallout shelters and related technologies promises to bring innovations that make these structures more effective, accessible, and sustainable. Advances in materials science, architecture, and emergency management are poised to revolutionize how we think about fallout protection, addressing many of the challenges faced by current designs. This chapter explores the potential innovations on the horizon for fallout protection, including advancements in shelter construction, materials, technology integration, and emergency planning.

Advanced Materials for Fallout Shelters

High-Strength, Lightweight Materials:

Composite Materials: Advances in materials science are leading to the development of high-strength, lightweight composite materials that could significantly improve the construction of fallout shelters. These materials, which combine the best properties of multiple substances, offer enhanced durability and protection while reducing the overall weight and cost of construction.

Graphene and Nanomaterials: Graphene, a material made from a single layer of carbon atoms arranged in a honeycomb lattice, is one of the strongest materials known. Its integration into shelter design could lead to structures that are incredibly strong, yet lightweight. Nanomaterials, with their ability to manipulate matter at the molecular level, could also provide new ways to create radiation-resistant coatings or enhance the structural integrity of shelters.

Radiation-Absorbing and Reflective Materials:

Radiation-Blocking Fabrics: Research is underway into fabrics and coatings that can block or absorb radiation, potentially allowing for new types of flexible, portable shelters. These materials could be used to line walls, ceilings, and floors, providing an additional layer of protection in traditional shelters or creating entirely new shelter designs.

Advanced Concrete: New formulations of concrete that incorporate materials like lead or specialized polymers can provide enhanced radiation protection without significantly increasing the weight or thickness of shelter walls. Self-healing concrete, which can repair cracks automatically, is another innovation that could improve the longevity and safety of fallout shelters.

Smart Shelters and Technology Integration

Smart Shelter Systems:

Automated Monitoring: The integration of smart technology into fallout shelters will allow for automated monitoring of radiation levels, air quality, and structural integrity. Sensors embedded in the shelter's walls can provide real-time data to occupants, helping them make informed decisions during a nuclear event.

AI and Machine Learning: Artificial intelligence (AI) and machine learning could be used to optimize shelter operations, such as managing energy use, ventilation, and water supply. AI could also assist in predicting the spread of fallout and adjusting shelter parameters to maximize safety and comfort.

Energy-Efficient Shelters:

Renewable Energy Integration: Future shelters could be equipped with renewable energy sources, such as solar panels, wind turbines, or geothermal systems, to ensure a continuous power supply during extended sheltering periods. These systems can be designed to work off-grid, making shelters more self-sufficient and reducing reliance on external power sources.

Battery Storage and Backup Power: Advances in battery technology, including solid-state batteries and other high-capacity energy storage systems, could provide reliable backup power for shelters. These innovations will be crucial for maintaining critical systems, such as air filtration and communication, during prolonged emergencies.

Advanced Air Filtration and Ventilation:

HEPA and Beyond: High-efficiency particulate air (HEPA) filters have long been the standard for removing radioactive particles from the air. Future shelters may use even more advanced filtration systems that incorporate nanotechnology or electrostatic precipitation to capture smaller particles and neutralize harmful gases.

Closed-Loop Ventilation Systems: Developing closed-loop ventilation systems that recycle and purify air within the shelter can help reduce the need for external air intake, further minimizing the risk of radiation exposure. These systems could also integrate with smart technology to adjust airflow and filtration based on real-time conditions.

Modular and Portable Shelter Designs

Modular Shelters:

Customizable Modules: Modular shelter designs allow for flexibility and scalability, enabling individuals or communities to build shelters that meet their specific needs. These shelters can be constructed from prefabricated modules that fit together like building blocks, making them easy to expand or reconfigure as circumstances change.

Ease of Assembly: Advances in construction techniques, such as 3D printing or robotic assembly, could make it possible to quickly deploy and assemble modular shelters in response to an imminent threat. These shelters could be pre-built and stored, ready to be assembled when needed, significantly reducing the time required to create a safe space.

Portable and Deployable Shelters:

Inflatable Shelters: Inflatable shelters, made from durable, radiation-resistant fabrics, could provide a portable option for fallout protection. These shelters could be stored compactly and quickly deployed in a crisis, offering a temporary refuge until more permanent shelter is available.

Pop-Up Shelters: Similar to inflatable shelters, pop-up shelters made from collapsible, lightweight materials could be easily transported and set up in a variety of locations. These shelters could be particularly useful for first responders or individuals in remote areas who need immediate protection.

Enhanced Emergency Planning and Communication

Real-Time Fallout Tracking and Prediction:

Satellite and Drone Technology: The use of satellites and drones to track and predict the spread of nuclear fallout in real-time could revolutionize emergency response. These technologies could provide accurate, up-to-date information on radiation levels, allowing individuals and authorities to make better decisions about evacuation and sheltering.

Predictive Modeling: Advanced computer modeling, using AI and machine learning, can predict the spread of fallout based on factors like wind speed, weather conditions, and topography. These models can help identify safe zones and optimize evacuation routes, improving survival rates and minimizing exposure.

Improved Communication Systems:

Mesh Networks: In the event of a nuclear strike, traditional communication networks may be disrupted. Mesh networks, which use decentralized nodes to create a resilient, self-healing communication system, could ensure that people remain connected even if central infrastructure fails. This technology could be integrated into shelters to maintain communication with the outside world.

Emergency Alert Apps: Mobile apps that provide real-time alerts, shelter locations, and emergency instructions could become increasingly sophisticated, offering personalized guidance based on the user's location and situation. These apps could integrate with smart shelter systems to provide tailored advice on when to shelter, evacuate, or take other protective actions.

Sustainable Shelter Design

Eco-Friendly Construction:

Recycled and Sustainable Materials: As sustainability becomes a greater priority, the use of recycled and eco-friendly materials in shelter construction is likely to increase. These materials could reduce the environmental impact of shelter construction and operation, making fallout shelters more sustainable in the long term.

Energy Efficiency: Future shelters will likely be designed with energy efficiency in mind, incorporating passive heating and cooling techniques, high-quality insulation, and energy-efficient appliances to minimize energy use and reduce reliance on external power sources.

Long-Term Survivability:

Self-Sustaining Shelters: The development of self-sustaining shelters, which can produce their own food, water, and energy, will be a key innovation in fallout protection. These shelters could include hydroponic or aquaponic systems for growing food, water purification systems that recycle wastewater, and renewable energy sources to ensure long-term survivability.

Water Harvesting and Recycling: Innovations in water harvesting, such as rainwater collection systems and atmospheric water generators, could provide a reliable source of clean water for shelters. Advanced filtration and recycling systems could ensure that water is used efficiently, reducing the need for large storage capacities.

Ethical and Social Considerations for Future Shelters

Universal Access and Affordability:

Inclusive Design: As fallout protection technology advances, there will be ethical imperatives to ensure that these innovations are accessible to all, not just the wealthy. Governments and NGOs may need to step in to subsidize or provide shelters for vulnerable populations, ensuring that everyone has access to protection.

Affordability: Efforts should be made to ensure that advanced shelter technologies remain affordable for the general public. This could involve government incentives, mass production techniques, or innovative financing models that allow more people to build or access shelters.

Community Integration:

Shared Shelters: Future shelters may be designed to accommodate entire communities, rather than just individuals or families. These shared shelters could foster a sense of community and cooperation, which is essential for long-term survival and recovery after a nuclear event.

Public-Private Partnerships: Collaboration between governments, private companies, and civil society could lead to the development of shelters that serve public needs while benefiting from private sector innovation and investment. These partnerships could help ensure that the most advanced shelter technologies are widely available.

The future of fallout protection is poised to be shaped by significant innovations in materials, technology, and shelter design. These advancements have the potential to make fallout shelters more effective, accessible, and sustainable, addressing many of the challenges faced by current designs.

As we look ahead, it's essential to consider not only the technical aspects of these innovations but also their ethical and social implications. Ensuring that all members of society have access to effective fallout protection, fostering community resilience, and promoting sustainability will be crucial as we develop the next generation of shelters. By embracing these innovations and addressing the associated challenges, we can create a safer and more resilient world in the face of nuclear threats.

Nuclear Winter: How to Survive a Long-Term Nuclear Winter

Surviving a long-term nuclear winter presents a daunting challenge, as the aftermath of a large-scale nuclear conflict could plunge the Earth into a period of drastically reduced temperatures, disrupted weather patterns, and widespread food shortages. Nuclear winter, a concept that emerged during the Cold War, refers to the severe global climatic cooling that could follow a nuclear war due to the massive amounts of soot and smoke released into the atmosphere, blocking sunlight. This chapter will provide a comprehensive guide to surviving a long-term nuclear winter, covering preparation, sheltering, food and water procurement, health considerations, and strategies for adapting to the harsh and prolonged conditions.

Understanding Nuclear Winter

The Causes and Effects:

Nuclear Detonations and Firestorms: The immediate aftermath of multiple nuclear detonations would result in large-scale fires, especially in urban areas, generating immense quantities of soot and smoke. These particles would be lifted into the stratosphere, where they could spread globally and remain for months or even years, reducing sunlight and lowering temperatures.

Global Cooling and Darkness: The reduced sunlight would cause a significant drop in global temperatures, leading to what is known as a "nuclear winter." This could result in a near-perpetual state of twilight, with daytime skies remaining darkened and cold, even during summer months.

Impact on Agriculture and Ecosystems: The cooling, combined with reduced sunlight, would severely disrupt global agriculture, leading to crop failures and food shortages. Ecosystems would be thrown into disarray, with many species unable to survive the sudden climate shift.

Preparing for Nuclear Winter

Stockpiling Supplies:

Food: A long-term nuclear winter would likely result in a global food crisis, so having a substantial stockpile of non-perishable food is crucial. Focus on foods that are calorie-dense, require minimal preparation, and have a long shelf life. Canned goods, dried beans, rice, pasta, freeze-dried meals, and grains are essential. Consider storing seeds for crops that can grow in low-light conditions, like certain root vegetables, for future food production.

Water: Access to clean water may be compromised during a nuclear winter due to contamination and the freezing of natural water sources. Store at least a year's supply of water, aiming for a minimum of one gallon per person per day. Water purification tools, such as filters, purification tablets, and boiling equipment, are also essential to ensure the safety of any collected water.

Heating and Fuel: With temperatures potentially plummeting, reliable heating sources are critical. Stockpile firewood, propane, or other long-lasting fuels. Consider investing in portable wood stoves or rocket stoves that can burn a variety of fuels efficiently. Alternative heating methods, such as solar heaters or thermal mass heating, can also help conserve fuel.

Shelter and Insulation:

Thermal Insulation: Your shelter must be well-insulated to retain heat and protect against the cold. This could involve adding extra layers of insulation to walls, sealing windows and doors to prevent drafts, and using thermal curtains or blankets to retain warmth. If possible, relocate to a basement or interior room where temperature fluctuations are less severe.

Backup Power and Lighting: With limited sunlight and the potential loss of grid power, having backup energy sources is vital. Solar panels, even with reduced efficiency in low light, can still provide some power, especially if paired with high-capacity batteries. Hand-crank generators, wind turbines, and fuel-based generators are also valuable. Stockpile candles, oil lamps, and battery-powered LED lights to maintain lighting.

Adapting to a Changed Environment

Food Production in Low-Light Conditions:

Indoor Gardening: Growing food indoors may become necessary during a nuclear winter. Hydroponic or aquaponic systems, which use water-based nutrient solutions instead of soil, can be set up indoors with artificial lighting. LED grow lights are energy-efficient and can support the growth of leafy greens, herbs, and other vegetables.

Mushroom Cultivation: Mushrooms, which thrive in dark, damp environments, can be a valuable food source in a nuclear winter. They can be grown in controlled indoor environments, providing essential nutrients and calories with relatively little effort.

Root Vegetables: Root vegetables like carrots, potatoes, and beets are more resilient to low light conditions and can be grown in containers indoors or in protected garden beds outside, provided they are insulated against the cold.

Hunting, Fishing, and Foraging:

Adapting to Scarcity: With wildlife populations potentially dwindling due to the harsh conditions, hunting, fishing, and foraging will be more challenging and require careful conservation of resources. Learn to trap small game, fish through ice, and identify edible plants, berries, and fungi that can survive the cold.

Preservation Techniques: Preserving food through drying, smoking, fermenting, or canning will be essential for long-term survival. Ensure you have the necessary equipment and knowledge to safely store food for extended periods.

Maintaining Health and Hygiene:

Health Care and Medical Supplies: Stockpile a comprehensive first aid kit and a year's supply of essential medications. Vitamins, especially Vitamin D, will be crucial due to the lack of sunlight. Learn basic medical skills to treat common ailments, injuries, and illnesses that may arise.

Mental Health: The psychological toll of living in a nuclear winter should not be underestimated. The extended darkness, isolation, and stress of survival can lead to depression, anxiety, and other mental health issues. Strategies for maintaining mental health include establishing a routine, staying connected with others, and engaging in activities that provide a sense of purpose and normalcy.

Sanitation: Maintaining sanitation in a nuclear winter is crucial to prevent the spread of disease. Stockpile hygiene supplies, such as soap, disinfectants, and sanitary products. Set up a waste management system that can operate without running water, such as composting toilets or bucket systems.

Surviving in a Community

Building a Support Network:

Mutual Aid: Surviving a nuclear winter will be easier with a support network. Form alliances with neighbors, family, and friends to share resources, knowledge, and labor. A group effort can lead to better defense, larger food production operations, and improved morale.

Community Shelters: Consider establishing or joining a community shelter where resources can be pooled, and efforts coordinated. A larger group can manage tasks like food production, defense, and medical care more effectively.

Security and Defense:

Protecting Resources: In a prolonged nuclear winter, scarcity could lead to desperation, making security a concern. Develop strategies for defending your shelter and resources against potential threats. This may include fortifying your shelter, creating a secure perimeter, and organizing a defense plan with your community.

Non-Violent Conflict Resolution: While security is important, non-violent conflict resolution within your community or group is equally critical. Establish clear rules, roles, and communication strategies to manage disagreements and maintain order.

Long-Term Planning and Adaptation

Preparing for the Long Haul:

Sustainability: Focus on developing systems that are sustainable for the long term. This includes renewable energy sources, regenerative food production, and conservation of supplies. The goal is to create a self-sustaining environment that can endure for months or even years.

Skills Development: Develop a broad range of survival skills that can help you adapt to changing circumstances. These might include blacksmithing, sewing, woodworking, and other practical skills that enable you to repair and maintain your shelter and tools.

Rebuilding and Recovery:

Reclaiming the Environment: As conditions eventually improve and sunlight returns, the focus will shift to reclaiming and restoring the environment. This might involve replanting crops, rebuilding infrastructure, and re-establishing communities. Having seeds, tools, and knowledge ready will be crucial for this stage.

Long-Term Survival: Surviving the initial years of a nuclear winter is only the first step. Long-term survival will require rebuilding food supplies, infrastructure, and society. Begin planning for this phase as soon as conditions allow, focusing on sustainability and community resilience.

Surviving a long-term nuclear winter requires extensive preparation, adaptability, and resilience. By understanding the potential challenges and taking proactive steps to prepare, individuals and communities can improve their chances of enduring such an extreme scenario.

The key to survival lies in comprehensive planning, stockpiling essential supplies, developing sustainable systems, and fostering strong community ties. While the prospect of a nuclear winter is terrifying, being well-prepared can make the difference between survival and disaster in the face of such a global catastrophe.

Nuclear Winter: Survival Techniques

When the unthinkable occurs, and the world is thrust into the bleak reality of a nuclear winter, survival will depend on your ability to adapt to a harsh, cold, and darkened environment. Nuclear winter, a phenomenon caused by the massive injection of soot and debris into the atmosphere following a large-scale nuclear conflict, could lead to a dramatic and prolonged drop in global temperatures. The sun's light would be blocked, crops would fail, and ecosystems would collapse, making day-to-day survival a formidable challenge.

In this chapter, we'll explore what you can expect during a nuclear winter and provide practical strategies for surviving in these extreme conditions. From growing food in low-light environments to keeping livestock alive, the key to survival lies in preparation, ingenuity, and a deep understanding of the new reality you'll face.

Understanding Nuclear Winter

The Onset and Duration:

Immediate Effects: After a series of nuclear detonations, large quantities of soot, smoke, and dust will be propelled into the upper atmosphere, where they can remain for months or even years. This layer of particulates will block sunlight, leading to a sharp drop in temperatures across the globe. Expect temperatures to fall rapidly, potentially by several degrees within days or weeks, creating an environment akin to a severe winter, even in typically temperate regions.

Extended Darkness: With sunlight severely diminished, daylight hours will be bleak, with the sky remaining a dark, overcast gray. The lack of light will disrupt natural circadian rhythms, making it difficult to determine day from night and impacting both human and animal behavior.

Long-Term Climate Impact: The effects of nuclear winter could last for years, with a slow recovery as particulates gradually settle out of the atmosphere. During this time, global agriculture would be crippled, water sources could freeze or become contaminated, and the overall ecosystem would be thrown into disarray.

Adapting to Cold and Low-Light Conditions

Maintaining Heat in Your Shelter:

Insulating Your Shelter: The first priority in a nuclear winter is maintaining warmth within your shelter. Insulate walls, windows, and doors with materials like foam boards, thermal blankets, and even thick layers of clothing or blankets. Seal any gaps that could allow drafts to enter.

Efficient Heating Sources: Use energy-efficient heating methods such as wood stoves, which can burn a variety of fuels, including wood, coal, and even biomass like dried leaves or animal dung. Consider constructing a thermal mass heater, which can retain and radiate heat for hours after the fire has gone out, conserving precious fuel.

Backup Heating Methods: Have backup heating options, such as propane heaters, portable solar heaters, or even improvised heat sources like candle-powered clay pot heaters. Stockpile plenty of fuel, as gathering wood or other resources will become increasingly difficult as the environment deteriorates.

Growing Food in Low-Light Conditions:

Indoor Gardening: With outdoor agriculture severely limited, indoor gardening becomes essential. Set up hydroponic or aquaponic systems, which allow you to grow food without soil, using nutrient-rich water. These systems can be compact and efficient, making them ideal for indoor use.

LED Grow Lights: Invest in energy-efficient LED grow lights to provide the necessary light spectrum for plant growth. While sunlight will be scarce, these lights can simulate the necessary conditions for photosynthesis. Focus on growing fast-growing, low-light crops like leafy greens, spinach, kale, and certain herbs, which can be harvested quickly and continuously.

Root Vegetables: Root vegetables such as potatoes, carrots, and beets can be grown in containers indoors or in well-insulated garden beds. These crops are more tolerant of low-light conditions and can provide a reliable food source during extended periods of darkness.

Keeping Livestock Alive:

Livestock Shelter: If you keep livestock, they will need well-insulated and heated shelters to survive the cold. Construct or retrofit barns, sheds, or other enclosures to provide warmth and protection from the elements. Use straw or other insulating materials to line the floors and walls of their shelter.

Feeding Livestock: Stockpile feed well in advance, as foraging or grazing will not be possible during a nuclear winter. Consider growing fodder indoors using hydroponic systems to supplement stored feed. Fodder crops like barley or wheatgrass can be grown quickly and provide nutritious feed for animals.

Water Management: Ensure your livestock has access to clean, unfrozen water. This may require using heated water troughs or regularly breaking ice on outdoor water sources. Consider setting up a water collection and purification system that can operate independently of external sources, such as rainwater harvesting combined with filtration and treatment systems.

Long-Term Survival Strategies

Preserving Food:

Food Preservation Techniques: With fresh food production limited, preserving what you have is critical. Use methods like canning, drying, fermenting, and smoking to extend the shelf life of your food supplies. Invest in vacuum sealers and mylar bags with oxygen absorbers to protect dried goods from moisture and spoilage.

Cold Storage: In colder climates, take advantage of natural refrigeration by creating cold storage areas outside or in unheated parts of your shelter. This can help preserve fresh produce, dairy, and meats for extended periods.

Adapting Your Diet:

Nutrient-Rich Foods: Focus on nutrient-dense foods that provide the most calories, vitamins, and minerals per serving. Beans, lentils, whole grains, nuts, and seeds should be staples of your diet, supplemented with whatever fresh produce and animal products you can produce or preserve.

Supplements: Consider stockpiling dietary supplements, especially Vitamin D, which will be in short supply due to the lack of sunlight. Multivitamins and other essential supplements can help prevent deficiencies during extended periods of limited food variety.

Mental and Physical Health:

Maintaining Morale: The psychological effects of nuclear winter, such as depression, anxiety, and despair, will be significant. Establish routines that include physical activity, social interaction (even within your household or community), and mental engagement through games, reading, or creative activities.

Exercise: Regular physical activity is vital for maintaining physical health and boosting mental well-being. Indoor exercises like calisthenics, yoga, or using small exercise equipment can help keep your body strong in confined spaces.

Community Support: If you are part of a larger group or community, work together to share resources, knowledge, and labor. Community projects, like shared gardens or communal meals, can provide a sense of purpose and help sustain morale during tough times.

Preparing for the Aftermath

Long-Term Resilience:

Reclaiming and Restoring Land: As the effects of nuclear winter begin to subside and sunlight returns, focus on reclaiming and restoring agricultural land. Start with hardy crops that can tolerate cooler temperatures and begin reestablishing food production systems.

Rebuilding Society: As conditions improve, the focus will shift to rebuilding communities and infrastructure. Begin planning for this phase early, gathering the tools, seeds, and knowledge needed to kick-start recovery efforts. Collaborate with others to pool resources and work together towards rebuilding a functional society.

Learning from History:

Historical Lessons: Study past events where communities faced extreme weather, food shortages, or other catastrophic conditions. Learning from history can provide valuable insights into how to organize, survive, and eventually thrive in the aftermath of a nuclear winter.

Surviving a nuclear winter requires adaptability, resourcefulness, and a deep understanding of the challenges posed by a cold, dark, and resource-scarce world. By preparing thoroughly and thinking creatively, you can increase your chances of not only surviving but also maintaining a level of stability and hope in the most challenging of circumstances.

Checklist: Items to have in a Nuclear War and the following Nuclear Winter

A comprehensive checklist is essential for ensuring you have all the necessary items to survive a nuclear war and the subsequent nuclear winter. This list covers the most critical supplies and tools to help you prepare for immediate survival during a nuclear event and the long-term challenges of living through a nuclear winter. The items are grouped into categories for ease of reference.

Shelter and Warmth

Shelter Location:

Secure, insulated space (basement, fallout shelter, or heavily reinforced interior room).

Insulation Materials:

Extra insulation for walls, windows, and doors (e.g., thermal blankets, foam boards).

Heating Sources:

Portable wood stove or rocket stove.

Firewood, coal, propane, or other long-lasting fuel sources.

Hand warmers and thermal blankets.

Sleeping Gear:

High-quality sleeping bags rated for sub-zero temperatures.

Thermal blankets.

Insulated sleeping pads.

Food and Water

Non-Perishable Food:

Canned goods (vegetables, fruits, meats, soups).

Dried foods (beans, rice, pasta, lentils).

Freeze-dried meals.

Grains (oats, wheat, barley).

Nuts, seeds, and dried fruits.

Powdered milk, protein powder, and meal replacement shakes.

Hardtack, crackers, and other long-lasting snacks.

Water Storage:

Minimum of one gallon of water per person per day (at least a year's supply).

Water containers (e.g., large storage barrels, portable water tanks).

Collapsible water containers for transport and storage.

Water Purification:

Water filters (e.g., gravity-fed, pump-style, or portable filters).

Water purification tablets or drops.

Boiling equipment (e.g., portable stove, metal pots).

Solar still or atmospheric water generator.

Health and Hygiene

First Aid Kit:

Bandages, gauze, adhesive tape.

Antiseptics (e.g., hydrogen peroxide, alcohol wipes).

Tweezers, scissors, splints.

Pain relievers, anti-inflammatory medications, antacids.

Prescription medications (at least a year's supply).

Vitamins, particularly Vitamin D.

Antibiotics (if available and appropriate).

Anti-radiation tablets (Potassium Iodide).

Sanitation Supplies:

Soap, hand sanitizer, and disinfectants.

Toothbrushes, toothpaste, and dental floss.

Feminine hygiene products.

Toilet paper, baby wipes, and sanitary wipes.

Portable toilet or bucket toilet with waste bags and biodegradable liner bags.

Bleach (for disinfection and water purification).

Personal Protective Equipment:

Respirators or N95 masks.

Goggles and gloves (for protection during outdoor exposure).

Radiation detection devices (Geiger counter or dosimeter).

Radiation-proof clothing (if available).

Communication and Information

Emergency Communication:

Battery-powered or hand-crank radio with NOAA weather band.

Two-way radios or walkie-talkies.

Satellite phone (if available).

Information Resources:

Printed survival manuals and guides (e.g., nuclear survival, first aid).

Local maps with key locations marked (shelters, water sources, etc.).

Writing materials (notebooks, pens, pencils).

Backup Power:

Portable solar charger.

Extra batteries (for all devices).

Power banks or portable battery packs.

Hand-crank generator.

Backup Generator with extra fuel

<u>Tools and Equipment</u>

Basic Tools:

Multi-tool or Swiss Army knife.

Axe, hatchet, or saw (for cutting wood and other materials).

Shovel and entrenching tool.

Hammer and nails.

Wrenches, screwdrivers, and pliers.

Duct tape and rope or paracord.

Cooking and Food Preparation:

Portable camp stove with fuel.

Cookware (pots, pans, utensils).

Manual can opener.

Fire-starting tools (matches, lighters, fire-steel)

Lighting:

Flashlights (LED) with extra batteries.

Lanterns (battery-powered or fuel-based).

Candles and candle holders.

Oil lamps and lamp oil.

Miscellaneous:

Heavy-duty trash bags (for waste management and waterproofing).

Tarp or plastic sheeting (for emergency shelter or insulation).

Ziploc bags and vacuum-sealed bags (for food storage and waterproofing items).

Sewing kit (for repairs).

<u>Clothing and Protection</u>

Warm Clothing:

Insulated jackets and pants.

Thermal underwear and base layers.

Wool socks and heavy-duty boots.

Gloves, hats, and scarves.

Rain and Wind Protection:

Waterproof outerwear (jackets, pants).

Windproof clothing layers.

Ponchos or rain suits.

Protective Gear:

Respiratory masks (N95 or better).

Radiation-proof or chemical-resistant clothing (if available).

Mental and Emotional Well-Being

Entertainment and Distraction:

Books, puzzles, and games (to pass the time and reduce stress).

Writing materials for journaling or creative expression.

Musical instruments (if portable).

Routine and Structure:

Calendar and planner (to maintain a sense of normalcy).

Religious or spiritual materials (e.g., Bible, Quran, meditation guides).

Group activities or projects to foster a sense of community and purpose.

Security and Defense

Personal Defense:

Firearms and ammunition (where legal and appropriate).

Non-lethal defense tools (pepper spray, stun guns).

Bladed weapons (knives, machetes).

Shelter Fortification:

Reinforcement materials (plywood, metal sheeting).

Locks, chains, and barriers for doors and windows.

Motion detectors or alarms.

Surveillance:

Binoculars or a spotting scope.

Security cameras (if possible to power).

Long-Term Sustainability

Gardening and Food Production:

Seeds for low-light crops and root vegetables.

Hydroponic or aquaponic systems (with necessary equipment).

Grow lights (LED) with power source.

Soil, compost, and fertilizers.

Water Collection:

Rainwater collection system.

Atmospheric water generator (if available).

Water barrels and storage tanks.

Animal Husbandry:

Small livestock, such as chickens or rabbits (if practical).

Feed and necessary supplies for livestock care.

This checklist serves as a comprehensive guide to the most important items needed for surviving a nuclear war and the ensuing nuclear winter. Careful planning and thorough preparation are essential to increase your chances of survival in such extreme conditions.

Don't miss out!

Visit the website below and you can sign up to receive emails whenever Andrew Parry publishes a new book. There's no charge and no obligation.

https://books2read.com/r/B-A-FROLC-ISWAF

BOOKS 2 READ

Connecting independent readers to independent writers.